Student Skill Guide to Accompany

# Modern Motorcycle Technology

Third Edition

Michael Ross

Australia • Brazil • Canada • Mexico • Singapore • United Kingdom • United States

***Student Skill Guide to Accompany Modern Motorcycle Technology,* Third Edition**

**Michael Ross**

SVP, GM Skills & Global Product Management: Dawn Gerrain

Product Director: Matthew Seeley

Product Team Manager: Erin Brennan

Senior Director, Development: Marah Bellegarde

Senior Product Development Manager: Larry Main

Senior Content Developer: Meaghan Tomaso

Product Assistant: Maria Garguilo

Vice President, Marketing Services: Jennifer Ann Baker

Marketing Director: Michele McTighe

Marketing Manager: Jonathon Sheehan

Senior Production Director: Wendy Troeger

Production Director: Andrew Crouth

Senior Content Project Manager: Cheri Plasse

Senior Art Director: Benjamin Gleeksman

Cover image(s):

1. © Shannon Kirk
2. © ManoAfrica/istockphoto.com
3. © styleTTT/istockphoto.com
4. © saicie/istockphoto.com
5. © stockphotomania/Shutterstock



ISBN: 978-1-305-49748-1

**Cengage**
200 Pier 4 Boulevard
Boston, MA 02210
USA

Cengage is a leading provider of customized learning solutions with employees residing in nearly 40 different countries and sales in more than 125 countries around the world. Find your local representative at: **www.cengage.com.**

To learn more about Cengage platforms and services, register or access your online learning solution, or purchase materials for your course, visit **www.cengage.com.**

**Notice to the Reader**

Publisher does not warrant or guarantee any of the products described herein or perform any independent analysis in connection with any of the product information contained herein. Publisher does not assume, and expressly disclaims, any obligation to obtain and include information other than that provided to it by the manufacturer. The reader is expressly warned to consider and adopt all safety precautions that might be indicated by the activities described herein and to avoid all potential hazards. By following the instructions contained herein, the reader willingly assumes all risks in connection with such instructions. The publisher makes no representations or warranties of any kind, including but not limited to, the warranties of fitness for particular purpose or merchantability, nor are any such representations implied with respect to the material set forth herein, and the publisher takes no responsibility with respect to such material. The publisher shall not be liable for any special, consequential, or exemplary damages resulting, in whole or part, from the readers' use of, or reliance upon, this material.

Printed in the United States of America
Print Number: 07 Print Year: 2021

# Contents

# Preface

The *Student Skill Guide to Accompany Modern Motorcycle Technology, Third Edition,* is designed to reinforce students' comprehension of the core textbook material and to guide them through inspection, diagnostic, and service/repair procedures in the lab.

Each *Skill Guide* chapter is related to the content in the respective chapter in *Modern Motorcycle Technology*. The chapters include theory-based Shop Assignments and performance-based Job Sheets: The Shop Assignments are knowledge assessments that can be completed in the classroom, shop, or as homework assignments, while the Job Sheets offer step-by-step guidelines, checkpoints, and questions for hands-on maintenance and repair activities.

*Note to instructors: Chapters 19 and 20 include Job Sheets that require some initial instructor preparation prior to the students' performance of the tasks. Refer to the Answer Key for the Student Skill Guide on the Instructor Resource CD and the Instructor Companion Website for Chapters 19 and 20 for prep instructions for these Job Sheets. To access the Instructor Companion Website online, go to login.cengage.com, and create an account or log into your existing account.*

CHAPTER 1

# Introduction to Modern Motorcycle Technology

## Shop Assignment 1-1

Name ______________________ Date __________ Instructor ______________

## Identifying Street Motorcycles

### Objective

After completing this activity sheet, you should be able to identify the differences and similarities among types of street motorcycles.

### Directions

From the following list, choose three types of street bikes, then compare and contrast their intended use and technology.

| | | | |
|---|---|---|---|
| Custom Cruiser | Dual-Purpose Sport | Hot-Rod Cruiser | Scooter |
| Sport-Touring | Touring | MUV | Electric |

1. List your three choices.
   a. ______________________
   b. ______________________
   c. ______________________

2. Compare the engines found in these three types of street bikes and list five similarities. (Example: two stroke, four stroke, large or small displacement, single, twin, four cylinder, electric, etc.)

   Similarities:
   a. ______________________
   b. ______________________
   c. ______________________
   d. ______________________
   e. ______________________

3. List five differences found among the engines in the three street bikes you chose.

   Differences:

   a. ______________________________
   b. ______________________________
   c. ______________________________
   d. ______________________________
   e. ______________________________

4. List three things these street bikes have in common. (Example: frame type, fairing, wheels, brakes, etc.)

   Similarities:

   a. ______________________________
   b. ______________________________
   c. ______________________________

5. List three differences found on the chassis of the three types of street bikes you chose.

   Differences:

   a. ______________________________
   b. ______________________________
   c. ______________________________

INSTRUCTOR VERIFICATION:

# Shop Assignment 1-2

Name ______________________ Date __________ Instructor ______________

## Identifying ATVs

### Objective

After completing this activity sheet, you should be able to identify the differences and similarities among types of ATVs.

### Directions

Answer the following questions.

1. Identify each type of ATV in the photos below. A list of ATV types has been provided. Write the type of ATV below each photo.

   Racing Four-Wheeler    Three-Wheeler    Touring/Camping    Utility Four-Wheeler

   Multipurpose Utility Vehicle

A

______________________

B

______________________

C

______________________

D

Courtesy of Yamaha Motor Corporation, USA

_______________________________________________

2. Compare the three types of ATVs by listing three similarities. (Example: two stroke, four stroke, disc brake equipped, two-wheel drive, four-wheel drive, etc.)

   Similarities:

   a. _______________________________________________

   b. _______________________________________________

   c. _______________________________________________

3. List three differences found on the ATVs you identified.

   Differences:

   a. _______________________________________________

   b. _______________________________________________

   c. _______________________________________________

INSTRUCTOR VERIFICATION:

# Shop Assignment 1-3

Name ______________________ Date __________ Instructor ______________

## Identifying Off-Road Motorcycles

### Objective

After completing this activity sheet, you should be able to identify the differences and similarities among types of off-road motorcycles.

### Directions

Answer the following questions.

1. Below you will find three photos of off-road motorcycles. Identify the motorcycle and write the name on the line below each photo. Consult the list of off-road bikes if needed.

   Dual Purpose    Enduro    Motocross    Trials

A    B    C

______________    ______________    ______________

2. Compare the three types of motorcycles by listing four similarities. (Example: two stroke, four stroke, disc brake equipped, single shock suspension, etc.)

   Similarities:

   a. ______________________________
   b. ______________________________
   c. ______________________________
   d. ______________________________

3. List three differences found among the three motorcycles you chose.

   Differences:

   a. __________________________________________

   b. __________________________________________

   c. __________________________________________

INSTRUCTOR VERIFICATION:

# Shop Assignment 1-4

Name ______________________ Date __________ Instructor ____________

## Creating a Five-Year Career Plan

### Objective

After completing this activity sheet, you should have created a career plan that will guide you for the next five years.

### Directions

Having a career plan can mean the difference between success and failure. Create a plan that outlines where you want to be in your career five years after graduation. Set goals for yourself and write down steps that will enable you to reach them. For instance, you might decide you want to own your own repair shop. Being a shop owner means more than just knowing how to repair motorcycles. Consider what you will need to know and how you will acquire that knowledge. Complete the following.

1. One year after graduation:

______________________________________________

______________________________________________

______________________________________________

2. Two years after graduation:

______________________________________________

______________________________________________

______________________________________________

3. Three years after graduation:

______________________________________________

______________________________________________

______________________________________________

4. Four years after graduation:

______________________________________________

______________________________________________

______________________________________________

5. Five years after graduation:

______________________________________________

______________________________________________

______________________________________________

INSTRUCTOR VERIFICATION:

# Shop Assignment 1-5

Name ______________________ Date __________ Instructor ____________

## Career Opportunities

### Objective

After completing this activity sheet, you should be able to identify career opportunities available to you.

### Directions

From the list below, choose three careers in the motorcycle field that appeal to you and then list them in order of preference. Remember to always choose a career that plays to your strengths. If you are a "people person" who enjoys a new challenge every day, consider a career as a service writer. If you work best in a structured environment, a setup technician may be more to your liking.

| | | | |
|---|---|---|---|
| Instructor | Lot Attendant | Parts Person | Race Team Crewman |
| Salesperson | Service Writer | Setup Tech | Technician Service Manager |

1. List your three career choices:
   a. ______________________
   b. ______________________
   c. ______________________

2. Describe your strengths and how the careers you chose are suited to your strengths.

   Career choice 1:

   ______________________

   ______________________

   ______________________

   Career choice 2:

   ______________________

   ______________________

   ______________________

   Career choice 3:

   ______________________

   ______________________

   ______________________

INSTRUCTOR VERIFICATION:

# Safety First

## Shop Assignment 2-1

Name ______________________________ Date __________ Instructor ______________

### Create a Shop-Familiarization Floor Plan

#### Objective

After completing this activity sheet, you should be able to identify the location of important safety-related items in the shop.

#### Directions

Draw a simple floor plan of your shop and label the location of the items below.

| | | | |
|---|---|---|---|
| CPR Station | Eye-Wash Station | Fire Extinguisher | Fuel Dump |
| MSDS | Oil Dump | Phone | Shop Towel Disposal |

INSTRUCTOR VERIFICATION:

# Shop Assignment 2-2

Name ______________________ Date __________ Instructor ____________

## Creating a Fire Evacuation Plan

### Objective

Having an evacuation plan can mean the difference between life and death when a fire occurs. Upon completion of this activity sheet, you should know the quickest route out of the facility in the event of a fire.

### Directions

Create your plan now so you know what to do should you ever need to evacuate. Your plan should show an evacuation route and where you should meet once outside of the building. Using your knowledge of the facility and its floor plan, draw a simple map showing where the nearest exit is located and the quickest route to get to it.

1. Draw the floor plan of your facility and show the route to the nearest exit.

2. Ask the instructor where your class should meet in case of an evacuation and write it below.

______________________________________________

INSTRUCTOR VERIFICATION:

# Shop Assignment 2-3

Name ______________________ Date __________ Instructor ____________

## Creating an Emergency Contact Address Form

### Objective

After completing this activity sheet, you should have a reference card that can be used if an emergency requires contacting the next of kin or a relative.

### Directions

Create a form that gives the contact information for a relative in the event of an emergency.

1. Person who should be contacted in the event of an emergency: ______________________

2. Address: ______________________________________________

3. Phone number: ______________________ Cell: ______________________

4. Relationship to you: ______________________

INSTRUCTOR VERIFICATION:

# Shop Assignment 2-4

Name ______________________________ Date __________ Instructor ______________

## Locating Tools and Chemicals

### Objective

After completing this activity sheet, you should be able to identify the location of special tools and chemicals commonly used in the shop.

### Directions

Draw a simple floor plan of your shop and label the location of the items below.

Commonly Used Chemicals  Fuel  Oil  Special Tools

INSTRUCTOR VERIFICATION:

# Shop Assignment 2-5

Name ______________________________ Date __________ Instructor ______________

## Fire Safety

### Objective

After completing this activity sheet, you should be able to identify the fire triangle, identify the four fire classes, and know the proper use of a fire extinguisher.

### Directions

Answer the following questions.

1. What three things make up the fire triangle?
   a. ______________________________
   b. ______________________________
   c. ______________________________

2. There are four classes of fire: Class A, Class B, Class C, and Class D. Identify the class of fire created by each material below by putting the letter of each type in the correct column. The first one is done for you.

   a. Paper
   b. Carburetor cleaner
   c. Sales receipts
   d. Dust covers
   e. Cleaning cloths
   f. Titanium shavings
   g. Wiring insulation
   h. Propane
   i. Oily rags
   j. Work aprons
   k. Magnesium
   l. Work area partitions
   m. Gasoline
   n. Electrical box

| Class A | Class B | Class C | Class D |
|---|---|---|---|
| a | ____ | ____ | ____ |
| ____ | ____ | ____ | ____ |
| ____ | ____ | ____ | ____ |
| ____ | ____ | ____ | ____ |
| ____ | ____ | ____ | ____ |
| ____ | ____ | ____ | ____ |
| ____ | | | |

3. List the type of extinguisher you would use to fight each type of fire by writing the letter of the extinguisher type next to the class. You may use some choices more than once.

   a. $CO_2$ b. Foam c. Halon d. Dry powder e. Water

   Class A: __________ Class B: __________ Class C: __________ Class D: __________

4. What is an incipient fire? *Circle the correct letter.*
   a. A fire that is less than 2 feet in area.
   b. A fire that is smoldering, but no flames are visible.
   c. A fire started by arson.

5. What does PASS stand for?
   P: ____________________
   A: ____________________
   S: ____________________
   S: ____________________

6. Should a fire extinguisher be aimed at the **base** or **flames** of a fire? *Circle one.*

7. Look at the fire extinguishers in your shop and list the class of fire(s) they will fight.
   ____________________ ____________________
   ____________________ ____________________

8. What symbol(s) is/are on the fire extinguisher(s) in your shop? ____________________

INSTRUCTOR VERIFICATION:

# Shop Assignment 2-6

Name ______________________________ Date __________ Instructor ______________

## Tool Safety

### Objective

After completing this activity sheet, you should be able to identify the types of tools commonly used in the shop and their safety precautions.

### Directions

Answer the following questions according to the guidelines set forth in the chapter on tool safety. (Please note that every scenario described is real.)

1. Christine is trying to remove a wheel bearing but the bearing remover is broken. Instead, she is using a screwdriver as a chisel. Is this wrong and, if so, what is the solution?

   ______________________________________________

   ______________________________________________

   ______________________________________________

2. Barney is using an impact driver and a regular socket to loosen a 17 mm nut. Is this wrong and, if so, what is the solution?

   ______________________________________________

   ______________________________________________

   ______________________________________________

3. Tom just bored a cylinder and needs to clean the metal chips off the bore table. The vacuum is broken, so he is using an air blower to clean off the table. Is this wrong and, if so, what is the solution?

   ______________________________________________

   ______________________________________________

   ______________________________________________

4. Juan is mounting a new license plate on a bike he just sold but he does not have a screwdriver. He asks Pedro to toss him one. Is this wrong and, if so, what is the solution?

   ______________________________________________

   ______________________________________________

   ______________________________________________

5. Sahid dropped his hammer in the oil pan. The handle is coated with oil but he has to get the steering head bearings out before closing time, so he keeps working with the hammer without wiping it clean. Is this wrong and, if so, what is the solution?

6. Rajan's cut-off wheel is just about worn out. The edges are frayed and it's past the minimum diameter recommended by the manufacturer. He only needs to cut once more to get a handlebar end and he will be done for the day. Is this wrong and, if so, what is the solution?

7. Julie finds that her drill motor will not turn unless she pushes the power cord into the drill with one hand while holding the drill with the other. Is this wrong and, if so, what is the solution?

8. Carlos needs to clean the chain on his customer's bike so he puts it on the centestand. He lets the bike idle in gear so he can wipe off the chain with a shop rag. Is this wrong and, if so, what is the solution?

9. Mael is inflating a knobby tire but the bead wire is damaged and it will not bead properly. He decides to overinflate the tire to make it bead. Is this wrong and, if so, what is the solution?

10. Ernesto is drilling out a hole in a steel handlebar. He is using a drill press and holding the bar with his hand in case the drill bit catches. Is this wrong and, if so, what is the solution?

INSTRUCTOR VERIFICATION:

# Shop Assignment 2-7

Name ______________________________ Date __________ Instructor ____________

## Using Personal Protective Equipment (PPE)

### Objective

After completing this activity sheet, you should be able to identify types of commonly used PPE and their safety precautions.

### Directions

Answer the following questions according to the guidelines set forth in the chapter on tool safety. In some of the scenarios described, you may find that there is more than one type of PPE needed to protect the technician. List all that are appropriate.

1. You need to paint a battery box damaged by battery acid. What type(s) of PPE should be used?

   ______________________________________________

2. Some service departments require that you watch out for yourself by wearing these at all times. What type(s) of PPE should be used?

   ______________________________________________

3. You are doing some arc welding. What type(s) of PPE should be used?

   ______________________________________________

4. You are using a high speed cutting bit while porting a cylinder head. What type(s) of PPE should be used?

   ______________________________________________

5. You are cleaning the brake dust out of a brake drum. What type(s) of PPE should be used?

   ______________________________________________

6. You are filling a battery with acid. What type(s) of PPE should be used?

   ______________________________________________

7. You are unloading motorcycle crates. What type(s) of PPE should be used?

   ______________________________________________

8. A battery explodes while on the charger and battery acid gets in your eyes. What type(s) of PPE should have been used?

   ______________________________________________

9. You are dyno testing a bike in the dynamometer room. What type(s) of PPE should be used?

___

10. You need to remove a carburetor from the car-cleaning solution where it is resting at the bottom. What type(s) of PPE should be used?

___

INSTRUCTOR VERIFICATION:

# Shop Assignment 2-8

Name ______________________ Date __________ Instructor ______________

## Safe Riding Practices

### Objective

After completing this activity sheet, you should be able to identify unsafe riding practices.

### Directions

Motorcycle technicians test ride many different types of motorcycles every day. Using your textbook as a resource, answer the following questions according to the guidelines set forth in the chapter on rider safety.

1. A 1970 CB350 comes in for repair and has the original tires. Should you test ride this vehicle?

   Yes No *Circle the answer.*

   Explain your answer: ______________________

   ______________________

2. The customer complains that his bike is making an odd noise at 6,000 rpm. You take it for a test ride but cannot hear the noise with your helmet on. Should you leave off your helmet this one time?

   Yes No *Circle the answer.*

   Explain your answer: ______________________

   ______________________

3. You just installed steel-braided brake lines and new race-compound brake pads. The brakes are now very powerful. Should you try a maximum braking during the test ride?

   Yes No *Circle the answer.*

   Explain your answer: ______________________

   ______________________

4. The service manager insists that you wear a leather coat and gloves when going for a test ride but its 96°F outside and your manager is away at lunch. Should you test ride without the coat and gloves?

   Yes No *Circle the answer.*

   Explain your answer: ______________________

   ______________________

INSTRUCTOR VERIFICATION:

# Tools

## Shop Assignment 3-1

Name ______________________ Date __________ Instructor __________

### Tool Knowledge Assessment

**Objective**

By completing this activity sheet, you should be able to demonstrate basic tool knowledge.

**Directions**

After reading the chapter on hand tools, answer the following questions to the best of your ability.

1. How does a Torx bit differ in appearance from a Phillips bit?

   ______________________

2. Why is the adjustable wrench not used more often in motorcycle repair?

   ______________________

   ______________________

3. What is a flare nut wrench used for?

   ______________________

4. Why use a 6-point wrench when a 12-point wrench is available?

   ______________________

   ______________________

5. What is a dead blow hammer used for?

   ______________________

6. What is the difference between a tap and a die?

7. What is a thread file used for?

8. What is a breaker bar used for?

9. When is a 12-point socket superior to a 6-point socket?

10. What are the three most common drive sizes for sockets?

11. Why should you avoid using a test light on solid-state electronics found in fuel-injected motorcycles and ATVs?

12. The software package that controls billing, inventory, parts ordering, and work allocation.

INSTRUCTOR VERIFICATION:

# Shop Assignment 3-2

Name ______________________________ Date __________ Instructor ______________

## Tools

### Objective

After completing this activity sheet, you should be able to identify commonly used hand tools.

### Directions

Using your textbook as a resource, answer the following questions.

1. Identify the wrenches pictured below by choosing from the following list. Write the name of the tool below each photo.

| | | | |
|---|---|---|---|
| Adjustable Wrench | Box-End Wrench | Combination Wrench | Flare Nut Wrench |
| Open-End Wrench | Socket Wrench | | |

A

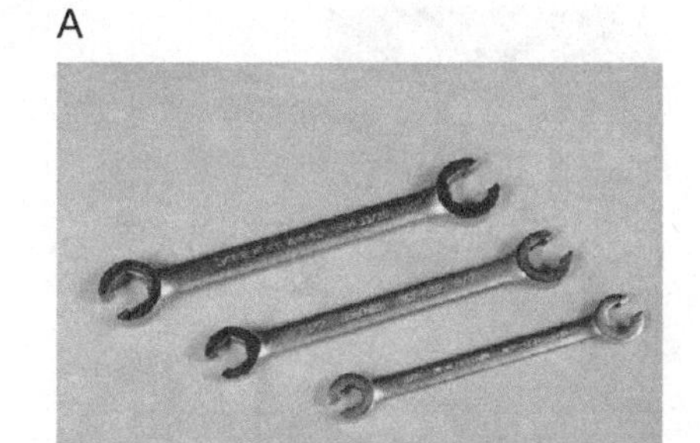

______________________

B

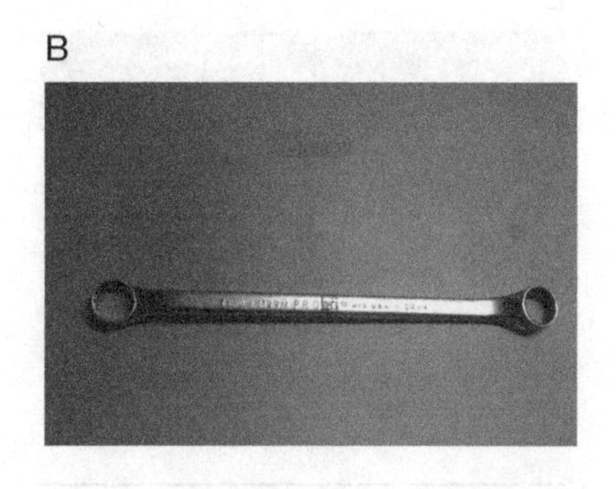

______________________

C

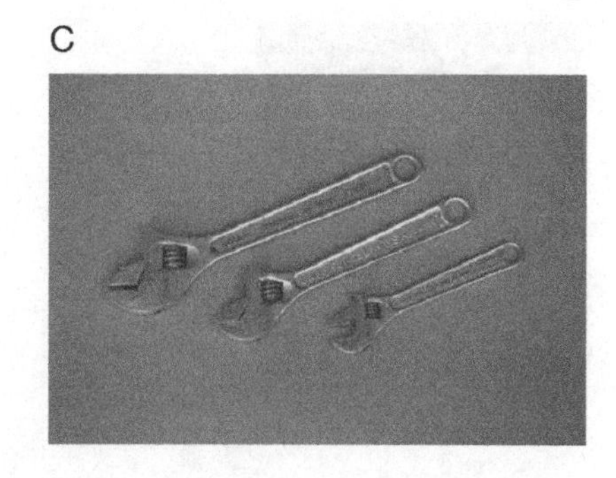

______________________

D

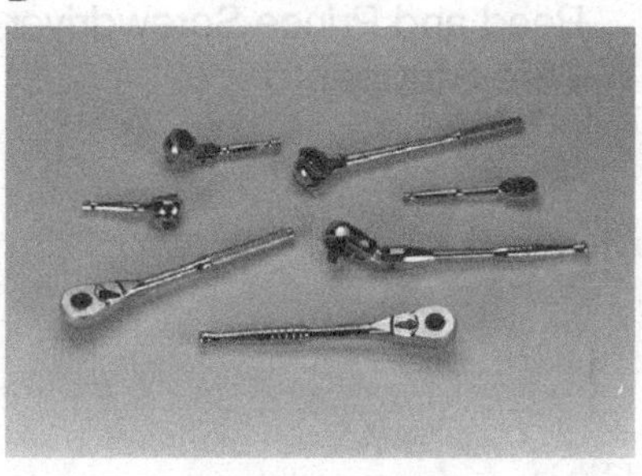

______________________

E

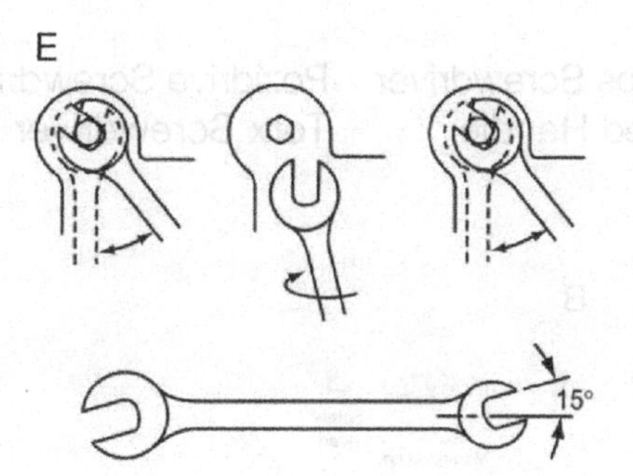


______________________

F

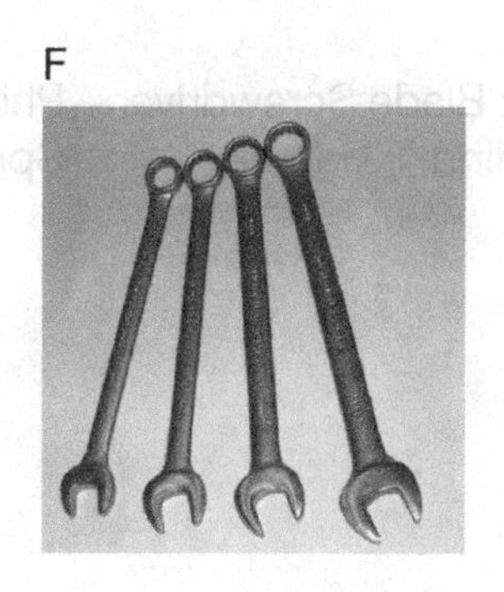

______________________

2. Identify the tools pictured below by choosing from the following list. Write the name of the tool below each photo.

| | | | |
|---|---|---|---|
| 6-Point Sockets | 6-Point Wrench | 12-Point Wrench | Hand-Type Socket |
| Hex Wrench | Impact Driver | Impact Socket | Ratchet Handle |

A

B

C

D

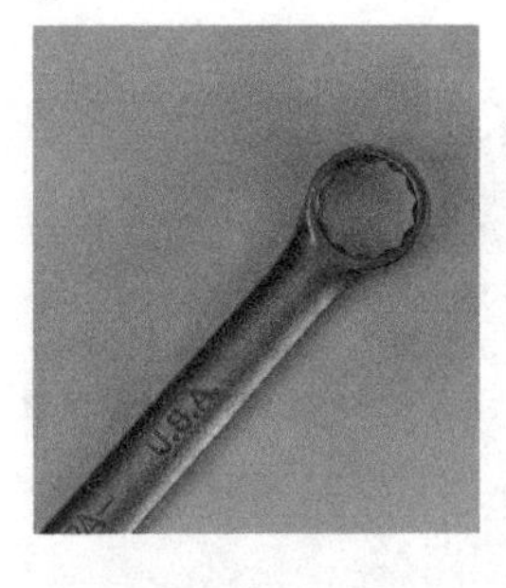

E

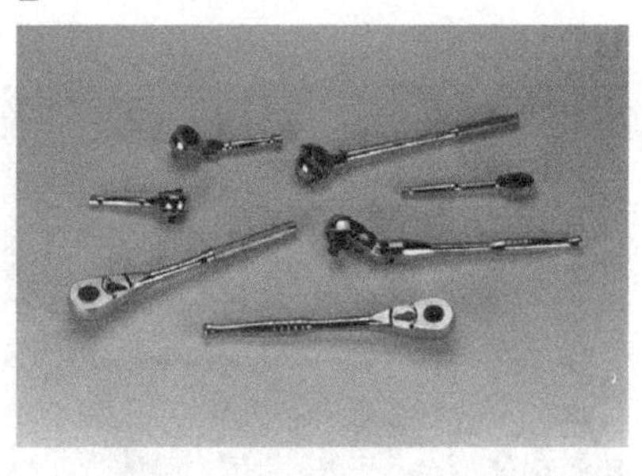

F

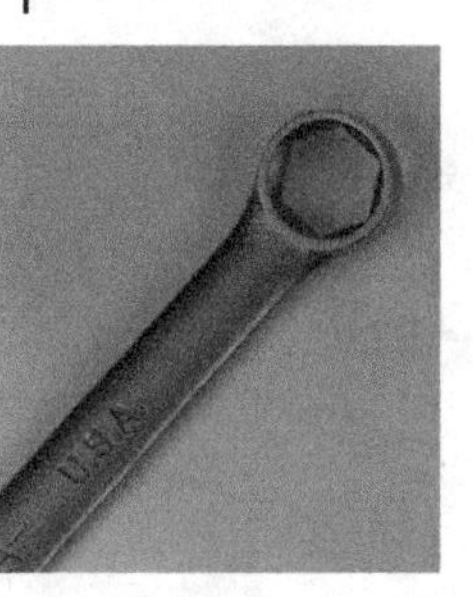

3. Identify the tools pictured below by choosing from the following list. Write the name of the tool below each photo.

| | | | |
|---|---|---|---|
| Flat Blade Screwdriver | Phillips Screwdriver | Pozidrive Screwdriver | Reed and Prince Screwdriver |
| Sliding T Handle | Speed Handle | Torx Screwdriver | Wire Stripper |

A

B

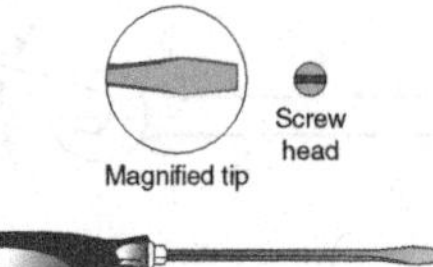


C

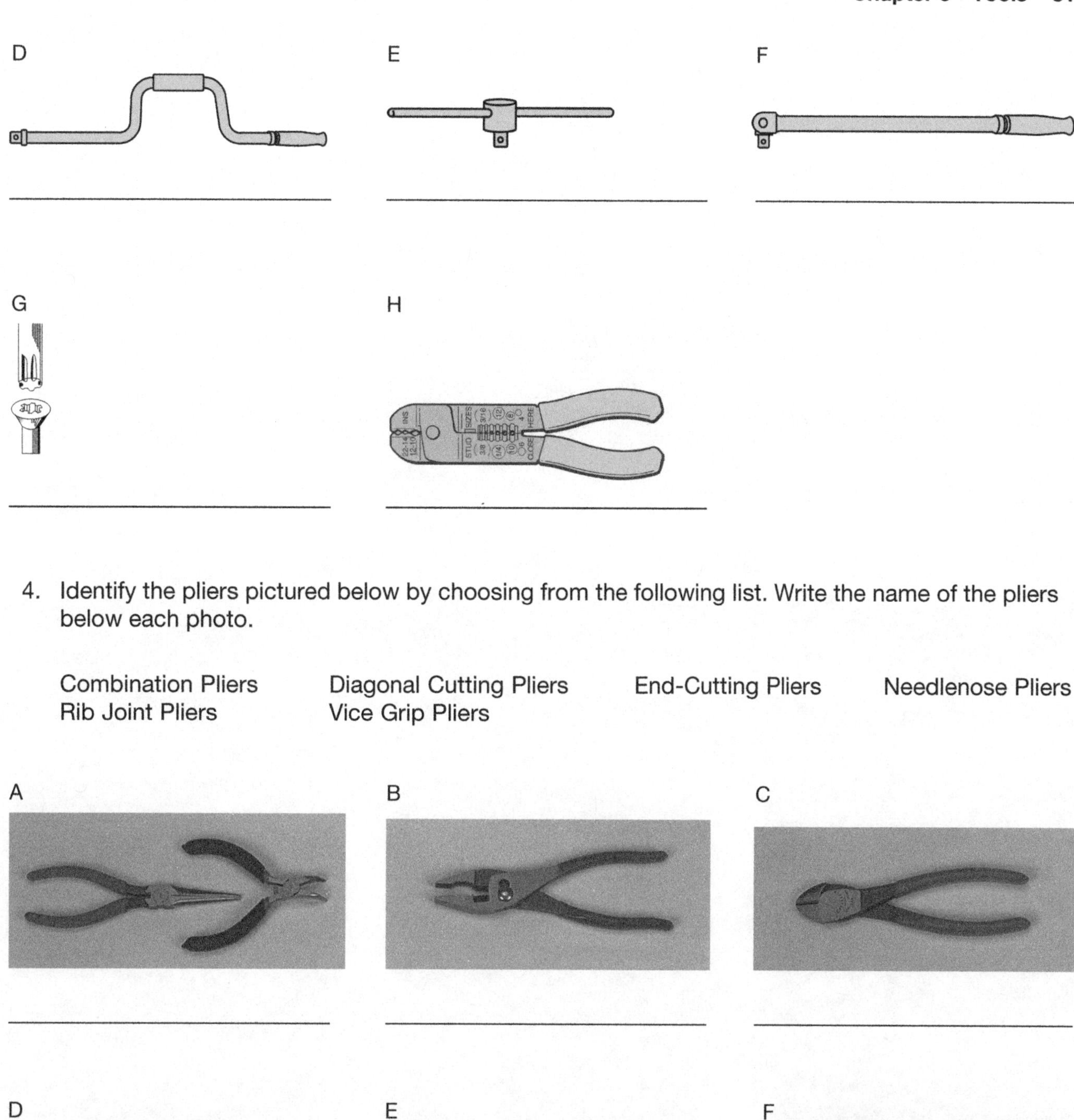


4. Identify the pliers pictured below by choosing from the following list. Write the name of the pliers below each photo.

Combination Pliers
Diagonal Cutting Pliers
End-Cutting Pliers
Needlenose Pliers
Rib Joint Pliers
Vice Grip Pliers

D

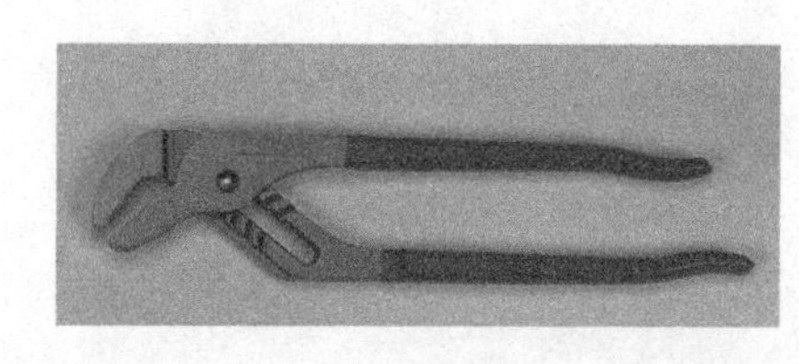

E

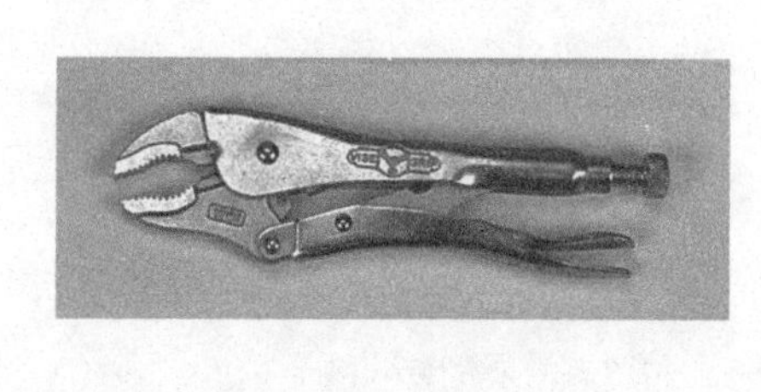

F

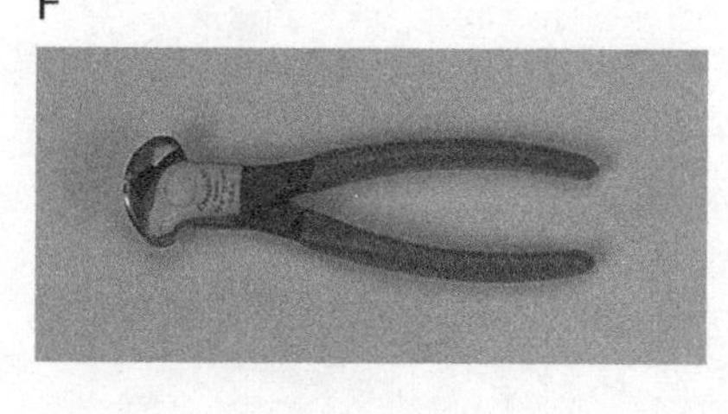

INSTRUCTOR VERIFICATION:

# Shop Assignment 3-3

Name ______________________ Date __________ Instructor __________

## Precision Measuring Tools

### Objective

After completing this activity sheet, you should be able to identify precision tools from photographs and drawings and know their use in the shop.

### Directions

Answer the following questions.

1. Name nine parts of a micrometer.

 ______________________________

2. Vernier calipers can make three types of measurements. What are they?

 ______________________________

3. Which is more accurate: a micrometer or a digital caliper?

 ______________________________

4. What is the dial indicator used to measure?

 ______________________________

5. Name two types of torque wrenches commonly found in the shop.

 ______________________________

6. What is a feeler gauge used to measure?

 ______________________________

7. Identify the precision measuring tools pictured below by choosing from the following list of names. Write the name of the tool below each photo.

| | | | |
|---|---|---|---|
| Dial Indicator | Digital Caliper | Feeler Gauge | Micrometer |
| Torque Wrench | Vernier Caliper | | |

A

B

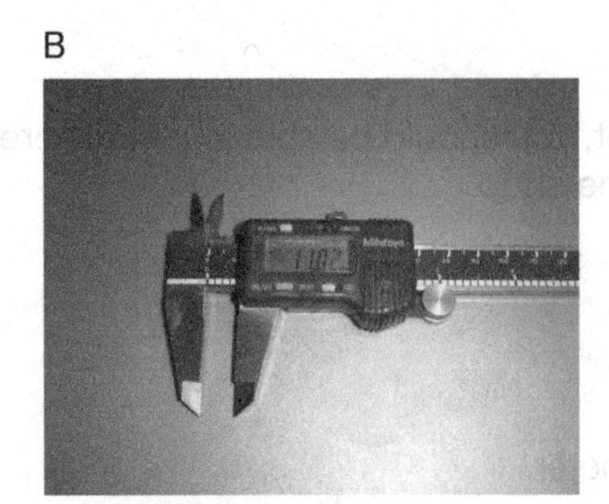

C

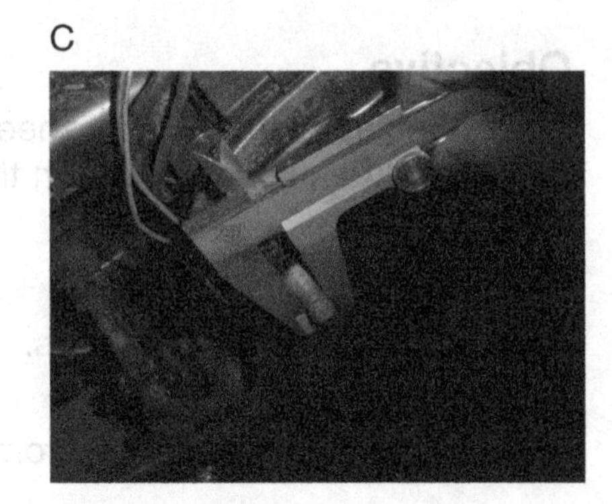

______________________ ______________________ ______________________

D

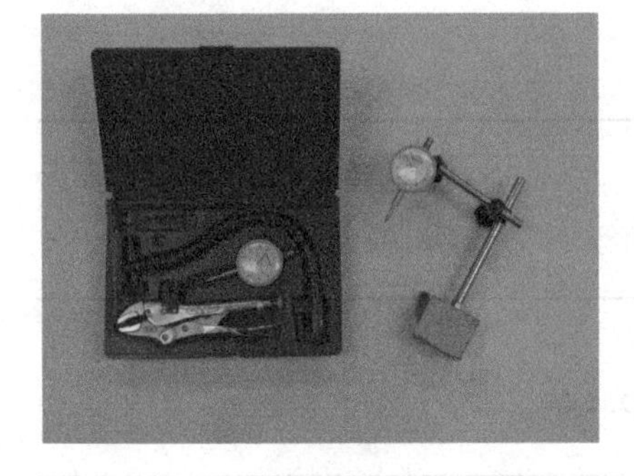

E

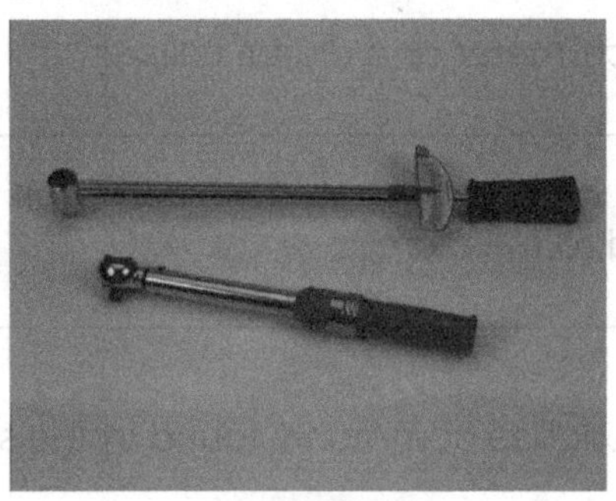

F

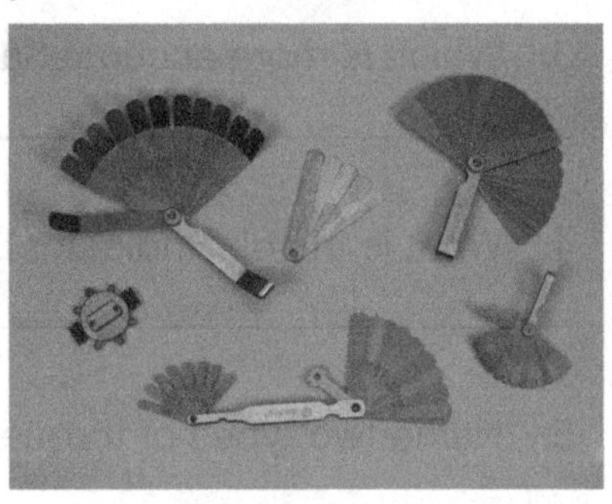

______________________ ______________________ ______________________

INSTRUCTOR VERIFICATION:

# Shop Assignment 3-4

Name ______________________________ Date __________ Instructor ____________

## Test Instruments

### Objective

After completing this activity sheet, you should be able to identify test instruments and know their use in the shop.

### Directions

Answer the following questions.

1. The most common electrical testing tool is a ____________________.

2. A timing light is used to check the motorcycle's ____________________ system.

3. What tool is used to check the cylinder pressure? ____________________

4. Among other things, this tool allows you to see how much unburned fuel is in the exhaust.
   ____________________

5. Use this tool to diagnose a fuel injection system that is "throwing a code."
   ____________________

| INSTRUCTOR VERIFICATION: |
| --- |

# Measuring Systems, Fasteners, and Thread Repair

## Shop Assignment 4-1

Name ____________________ Date __________ Instructor ______________

### Fastener, Measuring Systems, and Thread Repair Knowledge Assessment

#### Objective

After completing this activity sheet, you should be able to identify the two different measuring systems used on modern motorcycles and the fastener types used on these motorcycles.

#### Directions

Answer the following questions.

1. Name the two types of measuring systems used on modern motorcycles.
   a. ____________________ b. ____________________
2. What is the base point measurement of the Metric system?______________
3. How many millimeters are there in a meter?______________
4. What is the technical term used to describe the tension caused by tightening a fastener that holds two parts together? ______________
5. List the four critical bolt dimensions.
   a. __________ b. __________ c. __________ d. __________
6. If someone asked you to get a 10 mm bolt would you get a bolt that had a 10 mm **head** or a 10 mm **diameter**? *Circle your answer.*

7. You notice that a bolt head is stamped with the letter "L." What does this mean?

___

8. A bolt head is stamped with the number 8.8. Is this bolt **stronger** or **weaker** than a bolt stamped with the number 7? *Circle your answer.*

9. Why are castle nuts shaped as they are?

___

10. Technician A says that he often substitutes a stronger bolt than originally used by the manufacturer. Technician B says stronger bolts are brittle and may break if not used in the correct application. Who is right?

    A B *Circle the answer.*

11. What tool should you use to avoid overstretching a bolt when tightening it?

___

12. Where is a well nut used on a motorcycle? ___

13. Where is a cone nut used on a motorcycle? ___

14. When disassembling a component with various fastener sizes, which fasteners do you loosen first?

___

15. Lubricating fasteners is always a good idea.

    True False *Circle the answer.*

INSTRUCTOR VERIFICATION:

# Shop Assignment 4-2

Name ______________________ Date __________ Instructor ______________

## Fastener Anatomy and Identification

### Objective

After completing this activity sheet, you should know the anatomy of fastener types and be able to identify specific types of fasteners used on modern motorcycles.

### Directions

Answer the following questions.

1. Identify the basic parts of a fastener by marking up the drawing at the right. Label the location and dimensions of the fastener parts listed below

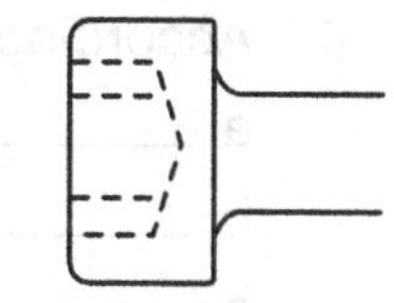

Head Length Nominal Diameter Thread Pitch

2. Write the type of bolt illustrated under its picture.

A

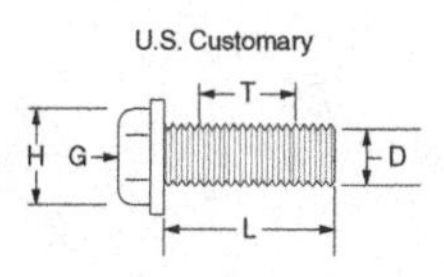


______________________

B

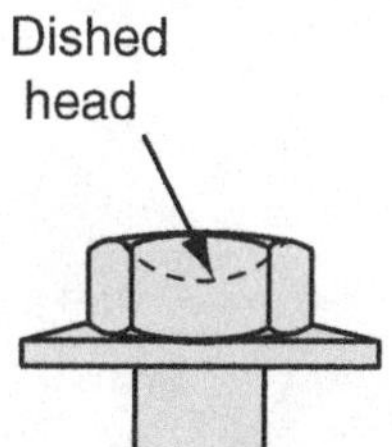


______________________

C

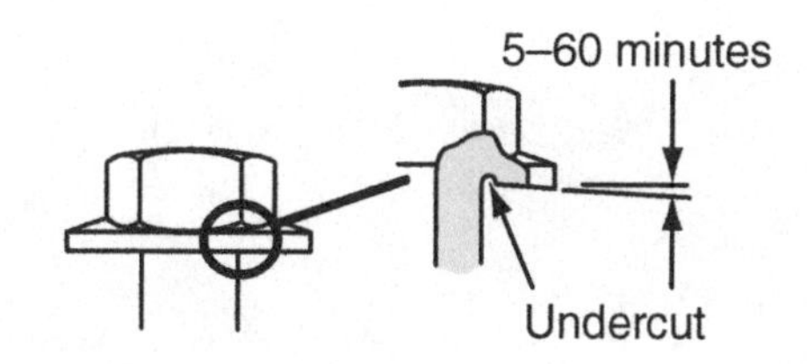


______________________

3. Write the type of nut illustrated under its picture.

D

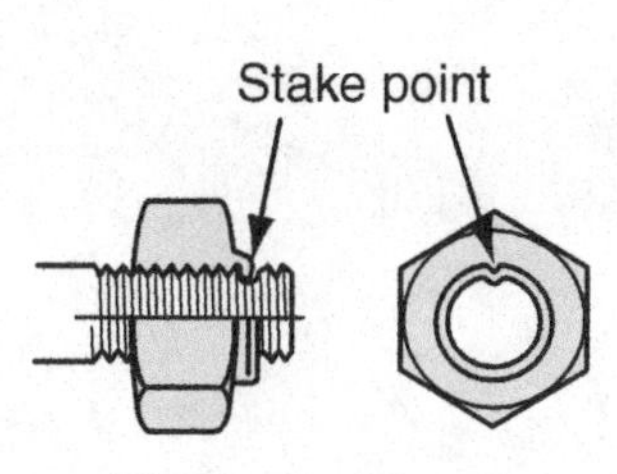


______________________

E

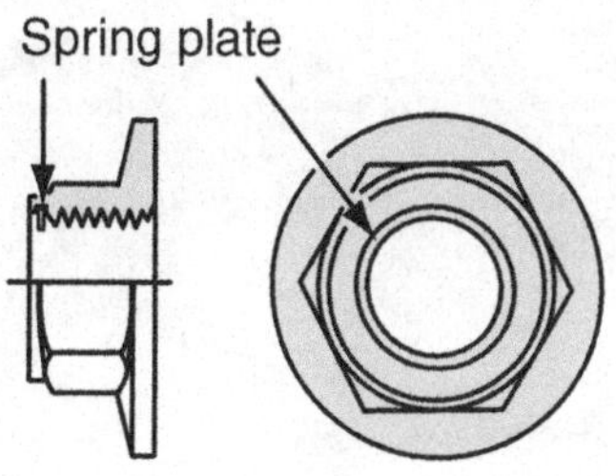


______________________

F

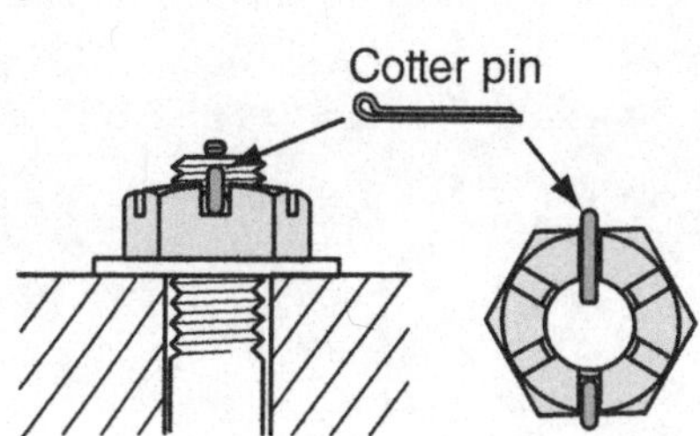


______________________

G

_______________

H

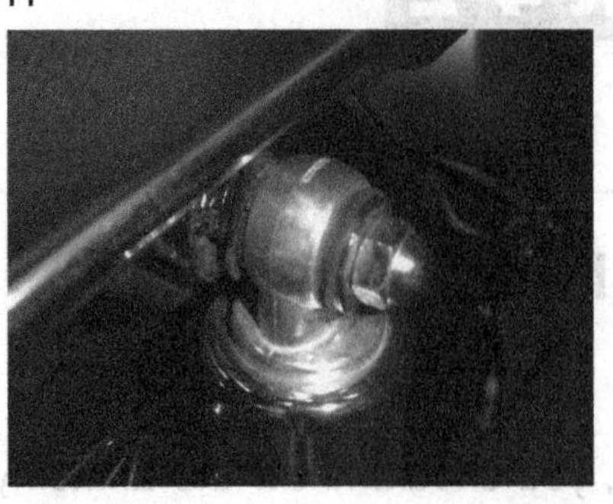

_______________

4. According to your textbook, where is a lock plate used? List three uses.
   a. _______________
   b. _______________
   c. _______________

INSTRUCTOR VERIFICATION:

# Shop Assignment 4-3

Name ______________________________ Date __________ Instructor ______________

## Torque Wrench Identification

### Objective

After completing this activity sheet, you should be able to identify the types of torque wrenches available to the technician and explain any precautions when using each type.

### Directions

Answer the following questions.

Write the type of torque wrench illustrated under its picture and note one precaution you should observe when using it.

A

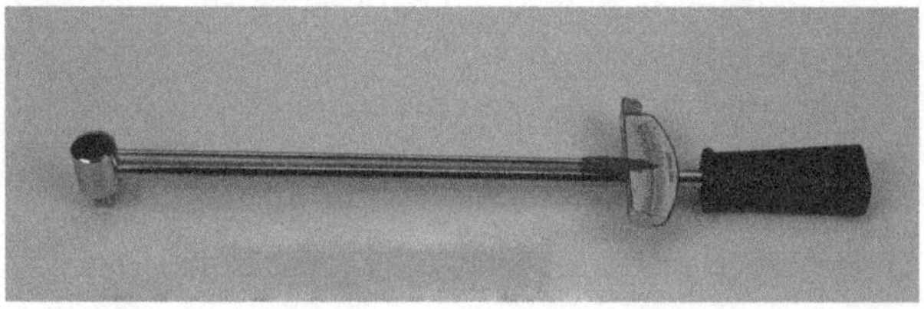

Precaution:

______________________________

______________________________

B

Precaution:

______________________________

______________________________

INSTRUCTOR VERIFICATION:

# Job Sheet 4-1

Name ______________________________ Date __________ Instructor ______________

## Using a Torque Wrench

### Objective

After completing this job sheet, you should be able to correctly use a torque wrench.

### Directions

Motorcycles often have assemblies held together with different size fasteners. Use the service manual to identify the amount of torque needed for each fastener and the order in which the fasteners are torqued.

### Tools and Equipment

Torque wrench, safety glasses, cylinder head or equivalent, service manual

1. Loosen all bolts in preparation for torquing. Check each one to make sure the threads are not damaged before starting.

2. Check whether the service manual specifies that the threads should be oiled.

3. List the sizes of fasteners and the amount of torque each size requires:
   a. ______________________________ ______________________________
   b. ______________________________ ______________________________
   c. ______________________________ ______________________________

INSTRUCTOR VERIFICATION:

# Job Sheet 4-2

Name ______________________________ Date __________ Instructor ______________

## Thread Repair

### Objective

After completing this job sheet, you should be able to remove broken fasteners and repair damaged threads.

### Directions

Your instructor will provide the necessary training aids to complete this lab sheet. You will purposefully overtorque a fastener until it breaks and then you will use a screw extractor to remove it.

### Tools and Equipment

6 mm bolt, aluminum block with threaded holes, screw extractor, drill motor, drill bit

1. Take a 6 mm bolt and thread it into one of the holes of the thread repair training aid. Tighten it with a wrench until it breaks off and then follow the procedure explained in the textbook (see pages 84–87) for using a screw extractor to remove it.

2. Show the broken bolt to the instructor before removing it.

3. Once the bolt is removed, have your instructor check your work.

| INSTRUCTOR VERIFICATION: |
| --- |

# Job Sheet 4-3

Name ______________________ Date __________ Instructor ______________

## Thread Insert

### Objective

After completing this job sheet, you should be able to properly install a thread insert. The thread insert will be installed so that it is flush with the top surface of the aluminum block and the insert will easily accept a new bolt. You will be able to torque the bolt to 8 ft lbs.

### Directions

Your instructor will provide the necessary training aids to complete this lab sheet. You will install a thread insert following the instructions provided with the thread insert kit.

### Tools and Equipment

Aluminum block with threaded holes, 6 mm thread insert kit, drill motor or drill press, tap handle

1. Follow the instructions provided with the thread insert kit and install it in the provided aluminum block.

2. After the thread insert is installed, thread a new bolt into the insert and torque it to 8 ft lbs.

INSTRUCTOR VERIFICATION:

# Basic Engine Operation and Configuration

## Shop Assignment 5-1

Name ______________________________ Date __________ Instructor __________

## Basic Engine Operation and Configuration Knowledge Assessment

### Objective

By correctly completing this assessment, you will demonstrate an understanding of basic two-stroke and four-stroke engine operation.

### Directions

Answer the following questions.

1. This engine gives a power stroke every 720 degrees. Is it a **two-stroke** or **four-stroke** engine? *Circle the answer.*

2. This engine has no poppet valves. Is it a **two-stroke** or **four-stroke** engine? *Circle the answer.*

3. This engine is environmentally friendly. Is it a **two-stroke** or **four-stroke** engine? *Circle the answer.*

4. The power stroke is an up stroke.
   True False *Circle the answer.*

5. This device measures an engine's torque and horsepower: ______________________

6. This engine uses the piston to time when the exhaust exits the combustion chamber. Is it a **two-stroke** or **four-stroke** engine? *Circle the answer.*

7. An engine's displacement is determined by the length of its stroke and the diameter of its piston.
   True False *Circle the answer.*

8. Torque is a measurement of an engine's ability to turn the back wheel.
   True  False  *Circle the answer.*

9. Compression ratio is a comparison of combustion chamber volumes at BDC and TDC.
   True  False  *Circle the answer.*

10. An engine with an 8.91:1 compression ratio will make more power than one with a 10.0:1 ratio.
    True  False  *Circle the answer.*

11. This is a crankshaft from a two-stroke engine. True False *Circle the answer.*

A

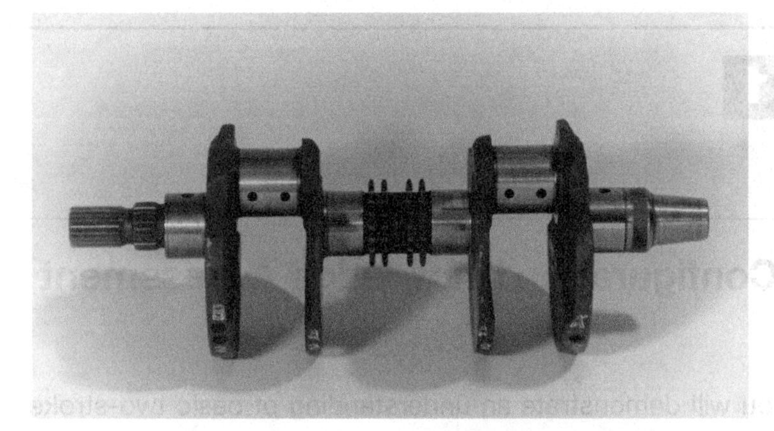

12. This is the connecting rod and piston from a four-stroke engine. True False *Circle the answer.*

B

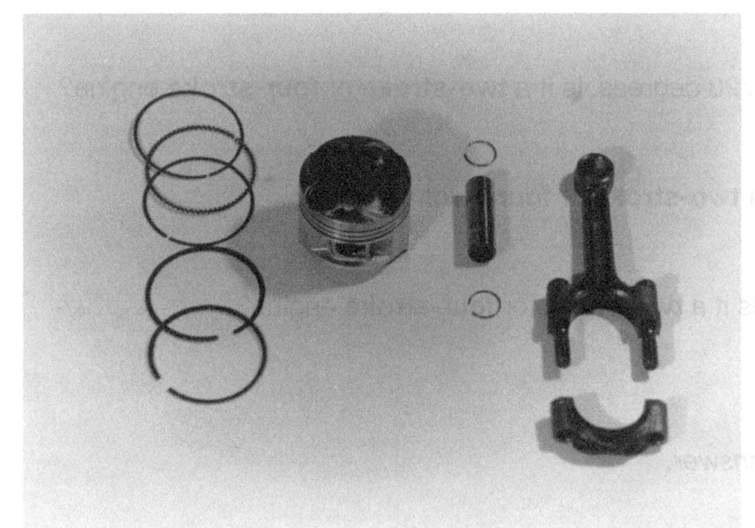

13. A 500cc engine has roughly the same displacement as a 45 ci in engine.
    True  False  *Circle the answer.*

14. Who invented the calculation for engine horsepower? ______________________

INSTRUCTOR VERIFICATION:

# Shop Assignment 5-2

Name ______________________ Date __________ Instructor ______________

## Identifying Engine Configurations

### Objective

After correctly completing this activity, you should be able to identify common types of motorcycle and ATV engine configurations.

### Directions

Answer the following questions. Write the type of engine configuration under each photo.

A

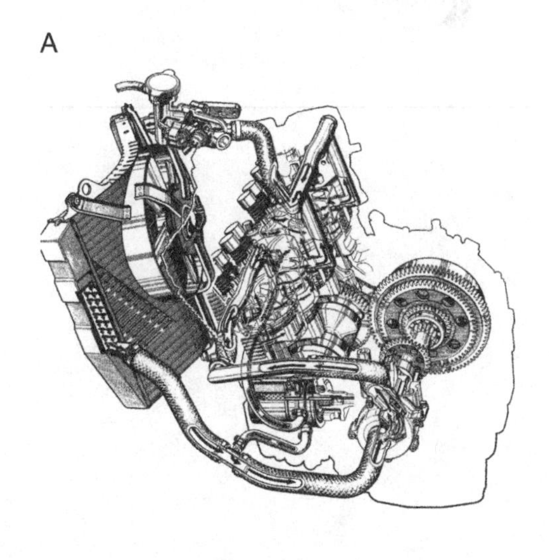

______________________

B

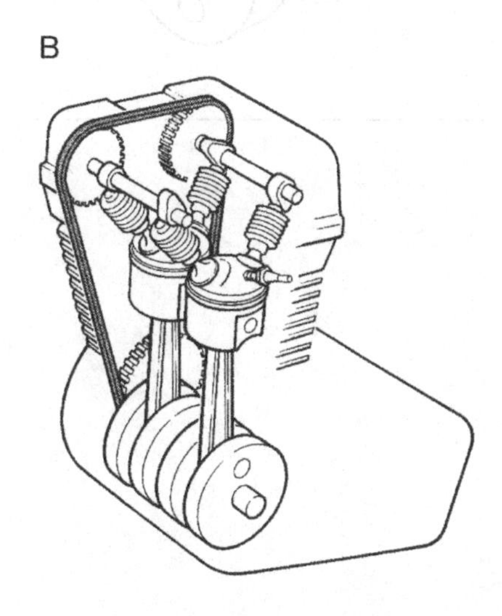

______________________

C

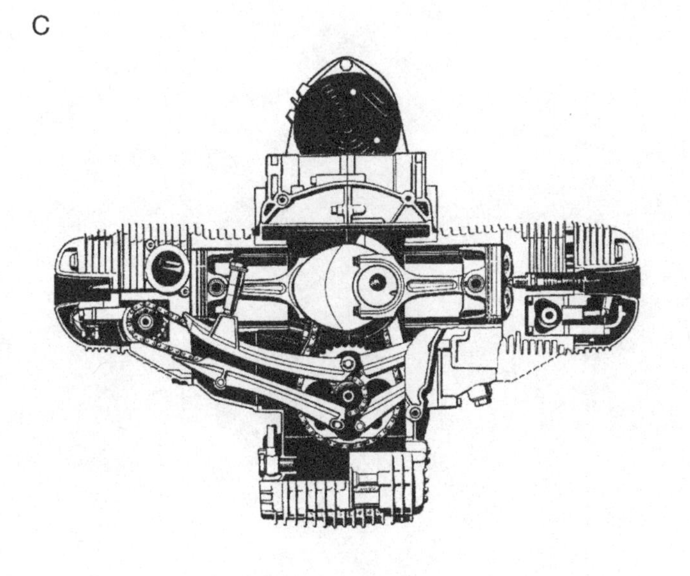

______________________

D

______________________

E

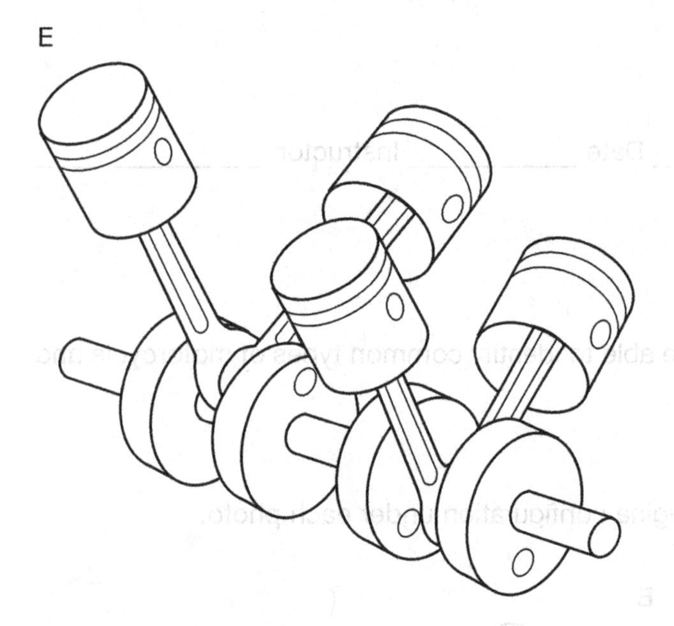

F

INSTRUCTOR VERIFICATION:

# Shop Assignment 5-3

Name ______________________________ Date __________ Instructor ______________

## Calculating Displacement

### Objective

After correctly completing this activity, you should be able to calculate engine displacement in both metric and SAE systems.

### Directions

Answer the following questions. Use this formula for displacement: $D = B \times B \times 0.7854 \times S \times N$.

1. A twin-cylinder engine has a bore of 52 mm and a stroke of 48 mm. Calculate its displacement.

   ______________________ cc

2. Use the same bore and stroke numbers as above but this time make it a four-cylinder engine.

   ______________________ cc

3. A single-cylinder engine has a stroke of 3.5 in. and a bore of 3.0 in. What is its displacement?

   ______________________ in.

INSTRUCTOR VERIFICATION:

# Internal Combustion Engines

## Shop Assignment 6-1

Name ________________________________ Date __________ Instructor ______________

## Internal Combustion Engine Knowledge Assessment

### Objective

After completing this assessment, you will have full knowledge of internal combustion engines.

### Directions

Answer the following questions.

1. According to your textbook, there are seven scientific terms and principles associated with the operation of an internal combustion engine. List them below.
   a. ____________________
   b. ____________________
   c. ____________________
   d. ____________________
   e. ____________________
   f. ____________________
   g. ____________________

2. Using the terms and principles you named in question 1, use a single word to describe how each of the seven principles or terms relates to an internal combustion engine. There may be more than one correct answer for some of them.
   a. Matter: ____________________
   b. Viscosity: ____________________
   c. Boyle's law: ____________________
   d. Pressure differences: ____________________

e. Momentum: ______

f. Laws of motion: ______

g. Energy: ______

3. Name the four stages of engine operation in order of occurrence.

a. ______

b. ______

c. ______

d. ______

4. What does a poppet valve do in a four-stroke engine?

______

5. What replaces the function of the poppet valve in a two-stroke engine?

______

6. What is stellite, and where is it used?

______

7. Refer to the drawing and name the parts of a valve.

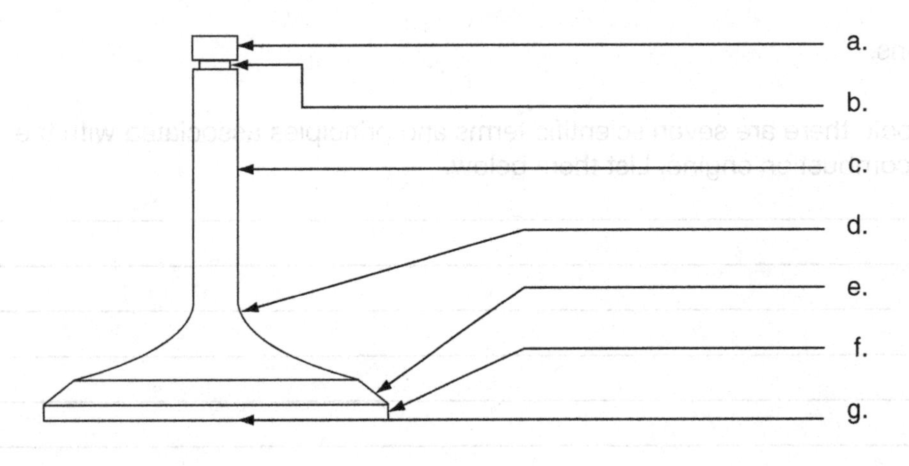


a. ______

b. ______

c. ______

d. ______

e. ______

f. ______

g. ______

8. Where does the valve in question 7 actually contact the combustion chamber?

______________________________

9. Name three common types of valve lash adjusters.
   a. ______________________________
   b. ______________________________
   c. ______________________________

10. What do the terms *ramp*, *dwell*, and *base circle* refer to?

______________________________

11. Name two ways a camshaft can be actuated.

______________________________

______________________________

______________________________

12. What term describes when both the intake and the exhaust valves are open at the same time?

______________________________

13. At what stage of engine operation does the action in question 12 occur?

______________________________

14. How many rings are there on a typical four-stroke piston?

______________________________

15. How many rings are there on a typical two-stroke piston?

______________________________

16. Where would you find a 120-degree crankshaft?

______________________________

17. At what stage does fuel induction occur?

______________________________

18. When does the spark plug fire?

______________________________

19. What does the transfer port do in a two-stroke engine?

______________________________

20. What does a labyrinth seal do on a two-stroke engine?

______________________________

21. Describe what is meant by an "offset" cylinder.

_______________

22. How does a cross plane crankshaft differ from a 180-degree crankshaft?

_______________

_______________

INSTRUCTOR VERIFICATION:

# Job Sheet 6-1

Name ______________________ Date __________ Instructor ______________

## Top-End Engine Inspection

### Objective

After completing this job sheet, you should be able to disassemble the top end of a two-stroke or four-stroke engine, identify its major components, and then reassemble it correctly.

### Directions

Using a training aid designated for this project by your instructor, consult the appropriate service manual for instructions on disassembling the engine's top end. After you have disassembled the top end, answer the questions and then reassemble the engine.

1. Does this engine have poppet valves?
   Yes No *Circle the answer.*

2. If you answered yes, how are the valves opened?
   Rocker arm Shim and bucket Hydraulic Does not apply *Circle the answer.*

3. How many rings are there on the piston?
   1 2 3 *Circle the answer.*

4. Does the wrist pin ride in a bearing?
   Yes No *Circle the answer.*

5. Does the piston have holes in the skirt?
   Yes No *Circle the answer.*

6. Does the combustion chamber have a squish band?
   Yes No *Circle the answer.*

7. Is the cylinder **cast iron** or **plated**? *Circle the answer.*

8. Is the cylinder crosshatched?
   Yes No *Circle the answer.*

9. What kind of top piston ring does the piston have?
   Standard Keystone Dykes Chrome plated *Circle the answer.*

10. Without removing the crankshaft, determine if it is a **single** or a **multipiece** unit.

    *Circle the answer*.

    Reassemble the engine according to the instructions in the service manual. Once the engine is reassembled, carefully turn the crankshaft two complete revolutions. If the engine will not turn past a certain point, stop immediately before you damage it. Consult your instructor for further instructions.

| INSTRUCTOR VERIFICATION: |
| --- |

CHAPTER 7

# Lubrication and Cooling Systems

## Shop Assignment 7-1

Name ______________________ Date ____________ Instructor ____________

## Lubrication and Cooling System Operation Knowledge Assessment

### Objective

After completing this assessment, you should be able to demonstrate your knowledge of cooling and lubrication system concepts and principles.

### Directions

Answer the following questions.

1. Oil performs four functions in a motorcycle engine. List them below.
   a. ______________________
   b. ______________________
   c. ______________________
   d. ______________________

2. API classifies oil by type. What is the highest API rating? ______________________

3. What does the "W" rating mean? ______________________

4. What is the viscosity index? ______________________

5. What API classification should be used in a motorcycle engine?

   SF SG SN *Circle all that apply.*

6. Name the two load types a ball bearing can carry.
   a. ______________________
   b. ______________________

7. Where are plain or precision insert bearings located in a motorcycle's engine?
______________________________

8. How does a two-stroke engine get its lubrication?
______________________________

9. What is the typical two-stroke premix ratio?
______________________________

10. What does *dry sump* mean?
______________________________

11. Name three types of oil pumps.
    a. ______________________________
    b. ______________________________
    c. ______________________________

12. When does the oil pressure relief valve operate?
______________________________

13. Name the three types of cooling systems.
    a. ______________________________
    b. ______________________________
    c. ______________________________

14. What does coolant at the "telltale" hole indicate?
______________________________

15. What is the checking pressure for a radiator cap? ______________________________

16. What is the recommended coolant-to-water ratio for motorcycles?
______________________________

17. How do you check the integrity of a cooling system? ______________________________

18. Which type of oil pump is most common on modern motorcycles?
______________________________

19. How do you test a thermostat? ______________________________

20. Are friction modifiers suitable for motorcycle clutches?
    Yes    No    *Circle the answer.*

21. Why would you use a hydrometer to check a cooling system?

_______________________________________________

22. Motorcycle manufacturers provide several different ways of checking the oil level. Name two of them.

a. _______________________________________________

b. _______________________________________________

INSTRUCTOR VERIFICATION:

# Shop Assignment 7-2

Name ______________________________ Date __________ Instructor ______________

## Calculating Premix Ratios

### Objective

After completing this activity, you should be able to mix fuel and oil at the correct ratio.

### Directions

Answer the following questions.

1. You are planning a big weekend of off-road riding, and you are bringing 10 gallons of gas. How much oil will you need to add to your gas for a 20:1 premix ratio?

   ______________________________________________

2. You have 1 gallon of gasoline, and your bike runs best on a premix ratio of 32:1. How much oil will you add to the gas?

   ______________________________________________

3. You are helping a friend at the track, and he needs to refuel for the next motocross. He has 3 gallons of gas left. How much oil should be added to the gas for a 40:1 ratio?

   ______________________________________________

4. You have 7.28 ounces of oil left and want to finish off the bottle. How much gas is needed to add to the oil for a 50:1 ratio?

   ______________________________________________

INSTRUCTOR VERIFICATION:

# Shop Assignment 7-3

Name ______________________________ Date __________ Instructor ______________

## Bearing Identification

### Objective

After completing this activity, you should be able to identify the different types of bearings and where they are most commonly used on a modern motorcycle.

### Directions

Identify the type of bearing pictured and where it might be used in a motorcycle.

A

Type of bearing:

______________________________

Applications:

______________________________

______________________________

______________________________

B

Type of bearing:

______________________________

Applications:

______________________________

______________________________

______________________________

C

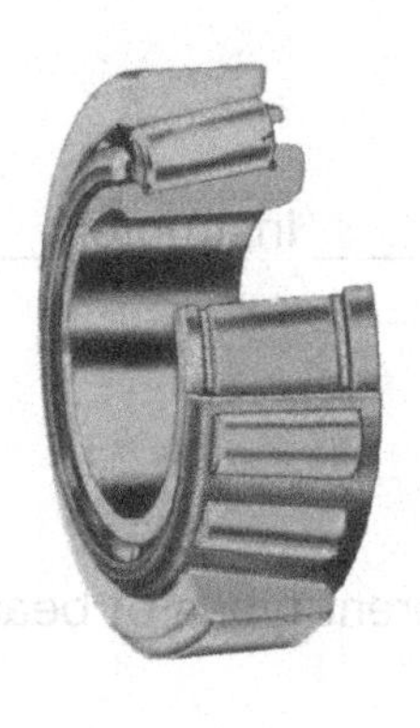

Type of bearing:

______________________________

Applications:

______________________________
______________________________
______________________________

D

Type of bearing:

______________________________

Applications:

______________________________
______________________________
______________________________

INSTRUCTOR VERIFICATION:

# Shop Assignment 7-4

Name ______________________ Date __________ Instructor ______________

## Lubricant Identification

### Objective

After completing this activity, you should be able to identify the different types of lubricants and their applications.

### Directions

Write down the type of oil that should be used in each vehicle listed below.

1. A high-performance sport bike: ______________________
2. A 1978 air-cooled four-cylinder bike ridden in Arizona: ______________________
3. A Kawasaki diesel Mule ATV: ______________________
4. A 2004 CR250 two-stroke engine: ______________________
5. A 2004 CR250 transmission: ______________________

INSTRUCTOR VERIFICATION:

# Job Sheet 7-1

Name ________________________________ Date __________ Instructor ______________

## Oil Pressure Check

### Objective

After completing this job sheet, you should be able to correctly test the oil pressure of an engine with plain bearings.

### Directions

Use the instructions provided in the appropriate service manual to check the oil pressure.

### Tools and Equipment

Oil pressure gauge, service manual

1. Using a motorcycle designated for this task by the instructor, remove the inspection plug and attach the oil pressure gauge.

2. Start the engine and observe the pressure reading.
   a. Observed oil pressure: ________________________________
   b. Service manual specification: ________________________________
   c. Did the engine meet the specification?

   Yes No *Circle the answer.*

3. Rev the engine up to 3,000 rpm and note any increase in pressure.
   Oil pressure at 3,000 rpm: ________________________________

4. What prevents the oil pressure from going too high when the engine revs up?
   ________________________________________________________________

5. Remove the oil pressure gauge and reinstall the inspection plug. Make the vehicle ready for inspection by the instructor.

INSTRUCTOR VERIFICATION:

# Job Sheet 7-2

Name ______________________________ Date __________ Instructor ____________

## Radiator Cap Inspection

### Objective

After completing this job sheet, you should be able to correctly test the radiator cap.

### Directions

Use the instructions provided in the appropriate service manual to check the radiator cap.

### Tools and Equipment

Cooling system pressure pump, cap adapter, service manual

1. A radiator cap must vent at the proper pressure to prevent damaging the cooling system. Obtain a pressure test kit from the instructor and note the pressure listed on the cap.
   Cap pressure: ______________________________

2. Attach the pressure pump to the cap and note the pressure at which the cap vents.
   Observed cap pressure: ______________________________

3. Did this cap pass the test?
   Yes No *Circle the answer.*

4. If a cap vents at too low a pressure, how will this affect the cooling system?
   ______________________________
   ______________________________

5. If you used a running vehicle for this test, replace the cap and make sure the cooling system is topped off. Prepare the bike for final inspection by the instructor.

INSTRUCTOR VERIFICATION:

# Job Sheet 7-3

Name ______________________ Date __________ Instructor ______________

## Cooling System Pressure Check

### Objective

After completing this job sheet, you should be able to correctly test the integrity of a cooling system.

### Directions

Use the instructions provided in the appropriate service manual to check the cooling system.

### Tools and Equipment

Cooling system pressure pump, radiator adapter, service manual

1. Remove the radiator cap from a cold vehicle and attach the pressure pump to the radiator.
2. What is the recommended test pressure for this vehicle? ______________________
3. Pump up the system to the recommended pressure and wait 1 minute. Check all hoses for leakage.
   a. Did the system hold pressure for 1 minute?

   Yes No *Circle the answer.*

   b. Did you note any leakage?

   Yes No *Circle the answer.*

   c. A system that does not hold pressure but shows no signs of external leakage may have a blown what? ______________________
4. Replace the radiator cap and prepare the vehicle for final inspection by the instructor.

INSTRUCTOR VERIFICATION:

# Job Sheet 7-4

Name ______________________________ Date __________ Instructor ______________

## Coolant Ratio Inspection

### Objective

After completing this job sheet, you should be able to correctly test the ratio of coolant.

### Directions

Obtain a hydrometer designated for this task by the instructor. Remove a sample of the coolant and note its strength.

### Tools and Equipment

Hydrometer

1. Remove the radiator cap from a cold engine and use the hydrometer to suck up a sample of the coolant.

2. Some hydrometers read in specific gravity, and others read in degrees of protection from freezing or boil over. What does your tester read?

   ______________________________________________

3. Does the coolant have sufficient ethylene glycol to protect it from freezing or boiling over?

   Yes No *Circle the answer.*

4. Replace the coolant you removed and reinstall the radiator cap. Prepare the vehicle for final inspection by the instructor.

INSTRUCTOR VERIFICATION:

# Fuel Systems

## Shop Assignment 8-1

Name ______________________________ Date __________ Instructor ______________

## Fuel Systems Knowledge Assessment

### Objective

After completing this assessment, you should be able to demonstrate your knowledge of fuel system concepts and principles.

### Directions

Answer the following questions.

1. A 110-octane fuel will give your engine more power.

   True  False  *Circle the answer.*

2. Detonation is dangerous to your engine.

   True  False  *Circle the answer.*

3. It is easy to hear detonation on a stock motorcycle.

   True  False  *Circle the answer.*

4. Technician A says that detonation is worse at higher altitudes. Technician B says that heavy loads, like climbing a hill, encourage detonation. Who is correct?

   A  B  *Circle the answer.*

5. More combustion heat equals more engine power.

   True  False  *Circle the answer.*

6. When speaking about air/fuel ratios, we are talking about pounds of fuel and pounds of air.

   True False *Circle the answer.*

7. The primary purpose of a carburetor is to vaporize fuel.

   True False *Circle the answer.*

8. If the fuel tank is higher than the carburetor, a fuel pump is needed.

   True False *Circle the answer.*

9. In California, the fuel tank must be vented to the atmosphere.

   True False *Circle the answer.*

10. The accelerator pump adds extra fuel when the carburetor is opened quickly.

    True False *Circle the answer.*

11. The air cutoff valve is part of the main circuit.

    True False *Circle the answer.*

12. Technician A says the active carburetor circuit is identified by the throttle position. Technician B says the active circuit is determined by engine rpm. Who is correct?

    A B *Circle the answer.*

13. Raising the jet needle makes the mid-range richer.

    True False *Circle the answer.*

14. The idle circuit will function perfectly if the main jet is removed.

    True False *Circle the answer.*

15. An EFI system that senses engine load by the use of a MAP sensor is referred to as a speed-density system.

    True False *Circle the answer.*

16. Fuel injection is rapidly replacing carburetors.

    True False *Circle the answer.*

17. Most fuel-injected bikes use direct injection.

    True False *Circle the answer.*

18. An injector is either off or on in normal operation.

    True False *Circle the answer.*

19. The ECM looks at the sensor readings and selects the correct map.

    True False *Circle the answer.*

20. The inside of a throttle body is shaped like a carburetor.

    True False *Circle the answer.*

21. An open loop means the fuel injection system needs an oxygen sensor.

    True False *Circle the answer.*

22. All modern street bikes have anti-tamper devices to prevent the owner from changing the idle mixture.

    True False *Circle the answer.*

23. Fuel-injected motorcycles make more power at high altitudes than carbureted bikes.

    True False *Circle the answer.*

24. Leaving the petcock open for a long period of time could cause the crankcase to fill with fuel.

    True False *Circle the answer.*

25. If your motorcycle detonates, it needs a higher octane fuel.

    True False *Circle the answer.*

26. Describe the difference between the motor octane number method and the research octane number method of measuring octane.

    ______________________________

    ______________________________

    ______________________________

    ______________________________

27. Which method of measuring octane provides the most realistic indication of how it will perform in motorcycle engines?

    ______________________________

28. What is meant by "throttle by wire"?

    ______________________________

    ______________________________

29. Name two ways you can adjust the fuel injection system on a motorcycle.

    1. ______________________________

    2. ______________________________

INSTRUCTOR VERIFICATION:

# Job Sheet 8-1

Name ______________________ Date __________ Instructor ______________

## Mechanical Slide Carburetor Inspection

### Objective

After completing this job sheet, you should be able to correctly identify all parts and passages in a slide-type carburetor and adjust the float level accurately.

### Directions

Use the instructions provided in the appropriate service manual to guide you through the inspection process.

### Tools and Equipment

Mechanical slide-type carburetor, service manual, float-level gauge, contact or carburetor cleaner, eye protection

1. Identify the carburetor by noting the identification (ID) number on the side. Write the ID number here: ______________________

2. Remove the float bowl, float, and the float valve.

3. Note the size of the main jet and the pilot jet. Write the sizes here.
   a. MJ: ______________________
   b. PJ: ______________________

4. According to the service manual, does this carburetor have the correct jets?
   Yes No *Circle the answer.*

5. Does this carburetor have an accelerator pump?
   Yes No *Circle the answer.*

6. Remove the carburetor top and pull out the slide.

7. Is the jet needle adjustable?
   Yes No *Circle the answer.*

8. In what position from the top is the clip? ______________________

9. Is this the correct position as noted in the service manual?
   Yes No *Circle the answer.*

10. Lightly seat the pilot screw. How many turns did it take to seat the screw? __________

11. What is the factory specification for pilot screw turns? ____________________

12. Squirt contact cleaner into the hole that the pilot jet screws into.

13. Where did the contact cleaner exit? ____________________

14. Squirt contact cleaner into the hole for the main jet.

15. Where did the contact cleaner exit? ____________________

16. Replace the jets and float assembly.

17. Hold the carburetor at eye level with the float hanging down. Push the float up gently until the valve seats. Note that there is a spring-loaded plunger in the valve that will compress after the valve seats. Using your float-level gauge, note how many millimeters the float hangs down with the valve just seated. (Plunger touching but not compressed.) ____________ mm

18. Some floats can be adjusted by bending the small tang that pushes the float valve. If this is the type you have, adjust the level to the service manual's specification.

19. Replace the jets and call your instructor over to check your float-level adjustment.

20. Replace the slide and carburetor top.

21. What kind of enrichment device does this carburetor have?

    Choke plate    Tickler    Enrichment circuit    *Circle the answer.*

22. Will lowering the clip on the jet needle make the mid-range **richer** or **leaner**? *Circle your answer.*

23. Replace the float bowl, being careful not to strip the screws.

| INSTRUCTOR VERIFICATION: |
|---|

# Job Sheet 8-2

Name ________________________________ Date __________ Instructor ____________

## CV Carburetor Inspection

### Objective

After completing this job sheet, you should be able to correctly identify the parts of a CV carburetor, check the float level, and describe how the carburetor operates.

### Directions

Use the instructions provided in the appropriate service manual to inspect this particular carburetor.

### Tools and Equipment

CV-type carburetor, service manual, float-level gauge, contact or carburetor cleaner, eye protection

1. Identify the carburetor by noting the identification number on the side. Write the ID number here:
   ____________________________________________
2. Remove the float bowl, float, and the float valve.
3. Note the size of the main jet and the pilot jet. Write the sizes here.
   a. MJ: ____________________________________________
   b. PJ: ____________________________________________
4. According to the service manual, does this carburetor have the correct jets?
   Yes No *Circle the answer.*
5. Some CV carburetors have three jets: a primary main, a main, and a pilot. Is this a three-jet carburetor?
   Yes No *Circle the answer.*
6. Check the float level according to instructions detailed in the service manual and adjust it if it is out of specification. Note: Some CV carburetors do not have adjustable floats.
7. Ask the instructor to check your float level.
8. Squirt contact cleaner into the hole for the pilot jet. Where did the contact cleaner exit?
   ____________________________________________
9. Does this carburetor have an air cutoff system?
   Yes No *Circle the answer.*

10. What circuit is the air cutoff system part of? ______________________________

11. Remove the slide and note the diaphragm. Is there a tab on the diaphragm to use to locate the slide in the carburetor?

    Yes        No        *Circle the answer.*

12. Turn the slide over. How many holes are in the bottom, not counting the hole for the jet needle?

    ______________________________

13. What do these holes do?

    ______________________________

14. If a diaphragm becomes torn, what effect will it have on the way the motorcycle runs?

    ______________________________

15. Replace all parts and ask the instructor for final verification.

| INSTRUCTOR VERIFICATION: |
| --- |

# Job Sheet 8-3

Name ______________________ Date __________ Instructor ______________

## Fuel Injection Component Location

### Objective

After completing this job sheet, you should be able to locate a fuel injection system on a fuel-injected motorcycle and correctly identify its parts.

### Directions

Use the appropriate service manual to locate the fuel injection sensors and components.

### Tools and Equipment

Fuel-injected motorcycle, service manual

1. Describe the location of the throttle body.

   ______________________________________________

2. Describe the location of the fuel injector.

   ______________________________________________

3. Describe the location of the MAP sensor.

   ______________________________________________

4. Describe the location of the coolant temperature sensor.

   ______________________________________________

5. Describe the location of the engine temperature sensor.

   ______________________________________________

6. Describe the location of the ECU.

   ______________________________________________

7. Describe the location of the cam position sensor.

   ______________________________________________

8. Describe the location of the speed sensor.

   ______________________________________________

9. Describe the location of the main relay.

   ______________________________________________

10. Describe the location of the malfunction light.

    ______________________________________________

| INSTRUCTOR VERIFICATION: |
| --- |

# Job Sheet 8-4

Name ______________________________ Date __________ Instructor __________

## Fuel Injection Diagnostics

### Objective

After completing this job sheet, you should be able to pull codes from the ECU's memory on a fuel-injected motorcycle and identify what component has failed.

### Directions

Use the instructions provided in the appropriate service manual to access the ECU memory.

### Tools and Equipment

Fuel-injected motorcycle, service manual

1. Read the section of the service manual that deals with pulling current and historic trouble codes and give a brief synopsis of the instructions below.

   ______________________________

   ______________________________

   ______________________________

   ______________________________

2. The malfunction light will blink out a trouble code when the ECU memory is accessed.
   a. What does a short blink represent?

      ______________________________

   b. What does a long blink represent?

      ______________________________

3. Does this vehicle have any current trouble codes?

   Yes No *Circle the answer.*

4. What are the historic trouble codes?

   ______________________________

5. Does this vehicle have any historic trouble codes in memory?

   Yes No *Circle the answer.*

6. What are the trouble codes?

______________________________________________________________

7. Replace any components you removed and prepare the vehicle for final inspection by the instructor.

| INSTRUCTOR VERIFICATION: |
|---|

CHAPTER 9

# Drives, Clutches, and Transmissions

## Shop Assignment 9-1

Name ______________________ Date __________ Instructor __________

## Clutch and Transmission Knowledge Assessment

### Objective

After completing this assessment, you should be able to demonstrate your knowledge of drive, clutch, and transmission principles.

### Directions

Answer the following questions.

1. Match the term with its definition.

| | | |
|---|---|---|
| a. Overall ratio | ______ | the ratio between the crankshaft and the transmission |
| b. Primary drive ratio | ______ | the ratio between the crankshaft and the rear wheel |
| c. Final drive ratio | ______ | the ratio between second gear and the gear it drives |
| d. Second gear ratio | ______ | the ratio between the countershaft sprocket and the rear wheel sprocket |
| e. Worm gear | ______ | a gear with angled teeth that creates a side force |
| f. Ring gear | ______ | a gear driven by a shank |
| g. Offset gear | ______ | two side-by-side spur gears |
| h. Bevel gear | ______ | a metal wheel with teeth |
| i. Helical gear | ______ | gear used to transmit power at 90 degrees |

2. Name the clutch component that is splined to the main shaft: ______________________

3. Name the clutch component that contains all of the other clutch components: ______________

4. This clutch component applies pressure to the clutch plates: ______________________

5. This clutch component is made from cork or Kevlar: ____________________

6. This clutch is cooled by air: ____________________

7. This clutch uses the engine's rpm to control engagement: ____________________

8. This clutch is usually found on scooters: ____________________

9. This clutch rotates in one direction only: ____________________

10. Your customer complains that she cannot accelerate quickly enough although the tachometer says the engine is speeding up. What may be wrong with the clutch?

____________________

11. Your customer complains that his bike is hard to shift in every gear and that he cannot find neutral when stopped at a light. What may be wrong with the clutch?

____________________

12. What would you check to troubleshoot the complaint in question 11?

____________________

13. Name the three types of transmission gears.

a. ____________________

b. ____________________

c. ____________________

14. Name the two types of constant mesh transmissions found on modern motorcycles.

a. ____________________

b. ____________________

15. This is the smallest gear on the mainshaft and it is fixed: ____________________

16. Top gear in this transmission is always 1:1: ____________________

17. Your customer complains that his transmission jumps out of second gear. What would you replace when repairing the problem? ____________________

18. You find that you can start a motorcycle by simply pulling in the clutch and kicking. What type of kick-start mechanism does this bike have? ____________________

19. Although the least efficient, this type of final drive system is the cleanest, strongest, and most trouble free. ____________________

20. In a constant mesh transmission, a fixed gear always mates with a ____________________ gear.

21. A CVT is most often found on this type of vehicle: ____________________

INSTRUCTOR VERIFICATION:

# Shop Assignment 9-2

Name ______________________________ Date __________ Instructor ______________

## Indirect Drive Power Flow

### Objective

After completing this job sheet, you should be able to correctly identify the power flow through an indirect drive transmission.

### Directions

Answer the following questions.

A

1. Using the photo above, identify the number of the gear using 1 for first gear, 2 for second gear, and so on. Identify the gear type by writing S for sliding gear, FW for freewheeling gear, and F for fixed gear next to each one.

2. Use a colored marker to show the power flow for first gear.

B

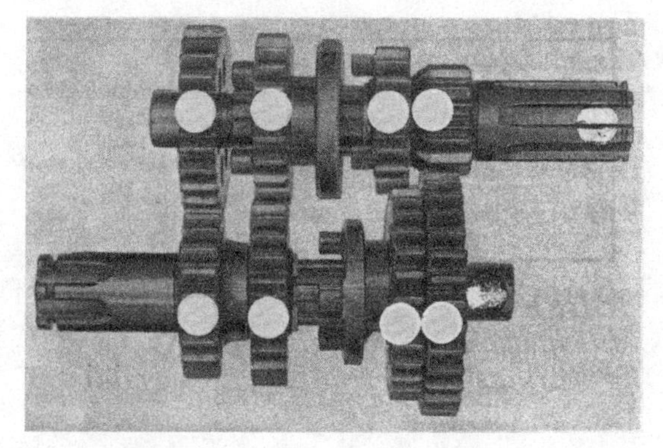

3. Use a colored marker to show the power flow for second gear in the photo above.

INSTRUCTOR VERIFICATION:

# Job Sheet 9-1

Name ________________________________________ Date ___________ Instructor _______________

## Drive Chain Adjustment

### Objective

After completing this job sheet, you should be able to correctly adjust the drive chain on a modern motorcycle.

### Directions

Using a motorcycle designated for this project, properly adjust the drive chain with the correct amount of slack.

### Tools and Equipment

Axle nut wrench, adjuster lock nut wrench, adjuster bolt wrench, service manual, steel rule

1. Place the motorcycle on its center stand or secure it to a stand so that the rear wheel is not touching the ground.

2. Loosen the axle nut and chain adjusters.

3. What is the service manual specification for this motorcycle?

   ________________________________________________________________________

4. Adjust the chain so that it has the correct amount of slack and the rear wheel is centered in the swing arm.

5. A motocross bike will have much more slack than a street bike when the rear wheel is suspended above the ground. Explain why.

   ________________________________________________________________________

6. Once you are satisfied with your work, tighten all fasteners and ask the instructor for a final checkoff.

INSTRUCTOR VERIFICATION:

# Job Sheet 9-2

Name ______________________ Date __________ Instructor ____________

## Clutch Inspection (Metric)

### Objective

After completing this job sheet, you should be able to correctly remove and replace the clutch in a modern motorcycle.

### Directions

Using a motorcycle designated for this project, remove the clutch plates, measure them, and replace them in the proper order.

### Tools and Equipment

Clutch cover bolt wrench, clutch hub nut wrench, clutch holder, Vernier calipers, flat plate, drain pan, clutch cover gasket

1. Using the service manual as your guide, remove the clutch cover.
2. With the clutch cover removed, use a clutch basket holder to secure the clutch while removing the clutch hub nut. This nut is usually very tight.
3. Remove the pressure plate and all clutch plates and friction discs.
4. Measure the friction discs and check their thickness against the service manual specification.
   a. Friction disc specification: ______________________ mm
   b. Friction disc measurement:
      1. ______________________
      2. ______________________
      3. ______________________
      4. ______________________
      5. ______________________
      6. ______________________
5. Lay the clutch plates on a flat surface. Did you find any that were warped? Ask the instructor to check that the plates are not warped.
   Yes No *Circle the answer.*
6. Measure the free length of the clutch springs:
   a. ________ b. ________ c. ________ d. ________ e. ________ f. ________

7. Reassemble the plates and discs in the correct order. Note that some bikes use different size plates in the clutch pack.

8. Ask the instructor to check your work before reinstalling the clutch cover.

INSTRUCTOR VERIFICATION:

# Job Sheet 9-3

Name ________________________________ Date __________ Instructor ____________

## Clutch Inspection (H-D)

### Objective

After completing this job sheet, you should be able to correctly remove and replace the clutch from a modern motorcycle.

### Directions

Using a motorcycle designated for this project, remove the clutch plates, measure them, and replace them in the proper order.

### Tools and Equipment

Clutch cover bolt wrench, clutch hub nut wrench, clutch holder, Vernier calipers, flat plate, drain pan, clutch cover gasket

1. Using the service manual as your guide, remove the clutch cover.

2. With the clutch cover removed, use a clutch basket holder to secure the clutch while removing the clutch hub nut. This nut is usually very tight.

3. Remove the pressure plate and all clutch plates and friction discs.

4. Measure the friction discs and check their thickness against the service manual specification.
   a. Friction disc specification: ______________________ mm
   b. Friction disc measurement:
      1. ______________________
      2. ______________________
      3. ______________________
      4. ______________________
      5. ______________________
      6. ______________________

5. Lay the clutch plates on a flat surface. Did you find any that were warped? Ask the instructor to check that the plates are not warped.

   Yes No *Circle the answer.*

6. Measure the free length of the diaphragm spring: ______________________________

7. Reassemble the plates and discs in the correct order.

8. Ask the instructor to check your work before reinstalling the clutch cover.

| INSTRUCTOR VERIFICATION: |
|---|

# Job Sheet 9-4

Name ________________________________ Date __________ Instructor ______________

## Final Drive Inspection

### Objective

After completing this job sheet, you should be able to correctly service the final drive unit of a shaft-driven motorcycle.

### Directions

Using a motorcycle designated for this project, remove the drain plug and drain the lubricant. Replace the lubricant with the correct amount and type.

### Tools and Equipment

Drain bolt wrench, drain pan, funnel

1. Drain the final drive fluid into a clean drain pan.
2. What type of lubricant is designated in the service manual for this type of vehicle?

   ________________________________________________________________

3. What is the recommended service interval for this machine?

   ________________________________________________________________

4. Look carefully at the fluid you just drained. If you find any metal filings, notify the instructor because it means there is a serious problem in the final drive unit.
5. Replace the drain plug, remove the fill plug, and refill the final drive.

INSTRUCTOR VERIFICATION:

# Job Sheet 9-5

Name ______________________________ Date __________ Instructor ____________

## Transmission Inspection

### Objective

After completing this job sheet, you should be able to correctly remove and replace a transmission in a modern motorcycle.

### Directions

Using a training aid designated for this project, remove the transmission and replace it so that it shifts through all gears when you are done.

### Tools and Equipment

Various hand tools, V-blocks

1. Remove the transmission assembly from the training aid and set the mainshaft on V-blocks so you can spin it.
2. Is the mainshaft straight?

   Yes No *Circle the answer.*
3. Carefully inspect the gears as you rotate the mainshaft. A slight nick or bent tooth will make a loud clunking noise when the transmission is put into use. Are there any bad teeth on any gears?

   Yes No *Circle the answer.*

   Which gear? ______________________________
4. Repeat the process for the countershaft.
5. Is the countershaft straight?

   Yes No *Circle the answer.*
6. Carefully inspect the gears as you rotate the countershaft. A slight nick or bent tooth will make a loud clunking noise when the transmission is put into use. Are there any bad teeth on any gears?

   Yes No *Circle the answer.*

   Which gear? ______________________________
7. Return the transmission assembly to the cases. Be sure that the shift drum and forks are in place and that they are correctly mated to the transmission shafts.
8. Verify that the transmission will shift through all gears. Note that you will have to hold the countershaft as you spin the mainshaft to get the dogs and slots to line up as you verify shifting.

| INSTRUCTOR VERIFICATION: |
|---|

CHAPTER 10

# Two-Stroke Engine Top-End Inspection

## Shop Assignment 10-1

Name ______________________ Date __________ Instructor ______________

## Two-Stroke Top-End Inspection Knowledge Assessment

### Objective

After completing this assessment, you should be able to demonstrate your knowledge of two-stroke engine inspection procedures.

### Directions

Answer the following questions.

1. Upon removing the piston from the training aid designated for this project, you notice tiny pins in the ring grooves. Why are they there?

   ______________________________________________

2. A seized two-stroke trail bike comes into the shop for repair and you are investigating why it seized. The owner says he used 50:1 premix. Is this the correct ratio for this engine?

   ______________________________________________

3. In how many places should the bore be checked for out of round and taper?

   ______________________________________________

4. Should the coolant be drained before removing the cylinder?

   ______________________________________________

5. What does the power valve do?

   ______________________________________________

6. What is the best resource for reassembling the power valve?

_______________

7. Can the power valve operation be checked while the engine is running?

_______________

8. How can you tell if a cylinder can be rebored?

_______________

9. Name two ways to clean the cylinder after it is rebored.

_______________

10. How are small scratches on the piston skirt dressed?

_______________

11. What does the small arrow stamped into the piston crown mean?

_______________

12. What would happen if a piston was installed backwards?

_______________

13. Is break-in important?

_______________

14. Can old piston rings be reused if they look fine?

_______________

15. How is a cylinder deglazed?

_______________

INSTRUCTOR VERIFICATION:

# Job Sheet 10-1

Name ______________________________ Date __________ Instructor ______________

## Two-Stroke Top-End Inspection

### Objective

After completing this job sheet, you should be able to correctly inspect the piston, cylinder, and reed valve of a two-stroke engine.

### Directions

Using a motorcycle designated for this project and the appropriate service manual, remove the head, cylinder, and piston. Measure the piston-to-cylinder clearance and correctly reassemble. Do this project in conjunction with 10-2.

### Tools and Equipment

Assorted hand tools, micrometer, dial bore gauge, service manual, feeler gauges, wrist pin remover

1. Loosen the cylinder head bolts using a crisscross pattern.
2. Remove the cylinder. Remember to remove the power valve linkage if the engine is so equipped.
3. Remove the wrist pin clip and slide the wrist pin out of the piston. Note: On engines that have been run, you may have to use a wrist pin remover.
4. Check the ring groove clearance with a feeler gauge.
   a. Manual specification for ring groove clearance: ______________________ mm
   b. Measured clearance: ______________________ mm
5. Remove the rings and square one up in the cylinder using the piston crown. Check the end gap.
   a. Manual specification for ring end gap: ______________________ mm
   b. Measured clearance: ______________________ mm
6. Check the bore size in six places. Write the bore sizes in the table below.

| *X*-axis | *Y*-axis | Out of Round |
|---|---|---|
| Top | | |
| Middle | | |
| Bottom | | |
| Taper | Taper | |

7. Does this cylinder meet the service manual's specifications for taper and out of round?

   Yes No *Circle the answer.*

8. Measure the piston size: ______________________________ mm

9. Calculate the piston-to-cylinder clearance:
   a. Top: ______________________________ mm
   b. Middle: ______________________________ mm
   c. Bottom: ______________________________ mm

10. Does this engine meet the specification for clearance? Use the largest number you calculated.

    Yes No *Circle the answer.*

11. Remove the reed assembly. Do the reeds touch the reed cage?

    Yes No *Circle the answer.*

12. Complete the next job sheet (10-2) before reassembling this engine.

| INSTRUCTOR VERIFICATION: |
|---|

# Job Sheet 10-2

Name ______________________ Date __________ Instructor ______________

## Power Valve Inspection

### Objective

After completing this job sheet, you should be able to correctly inspect the power valve of a two-stroke engine.

### Directions

Using a motorcycle designated for this project and the appropriate service manual, remove the power valve assembly, noting any damage or wear.

### Tools and Equipment

Assorted hand tools

1. Follow the instructions in the service manual and remove the power valve assembly.
2. Does this power valve system change the **exhaust port height**, the **exhaust chamber volume**, or both? *Circle the answer(s).*
3. According to the service manual, what is the service interval for the power valve?
   ______________________________________________
4. Reassemble the power valve.
5. Once the power valve has been reassembled, assemble the top end and make the training aid ready for final inspection.
6. Be prepared to demonstrate to the instructor how the power valve works and to explain how it broadens the power band.

INSTRUCTOR VERIFICATION:

# Two-Stroke Bottom-End Inspection

## Shop Assignment 11-1

Name ______________________________ Date __________ Instructor ______________

### Two-Stroke Bottom-End Inspection Knowledge Assessment

#### Objective

After completing this assessment, you should be able to demonstrate your knowledge of two-stroke engine inspection procedures.

#### Directions

Answer the following questions.

1. Crank seals should be replaced any time the crankshaft bearings are replaced.
   True  False  *Circle the answer.*

2. A leaky crank seal can result in a lean running condition.
   True  False  *Circle the answer.*

3. You must remove the top end first to service the crankshaft ______________________.

4. The clutch on a two-stroke engine seldom needs service.
   True  False  *Circle the answer.*

5. The transmission in a two-stroke engine is virtually identical to the transmission in a four-stroke engine.
   True  False  *Circle the answer.*

6. It is a waste of time to try to preserve the order of the gears and shims when removing them from the shaft.
   True  False  *Circle the answer.*

7. If the dogs on a slider gear are excessively worn, you should also replace the gear it mates with.

   True False *Circle the answer.*

8. Install thrust washers with the chamfered side facing the thrust load.

   True False *Circle the answer.*

9. Some engines require a case-splitter tool when splitting the cases.

   True False *Circle the answer.*

10. Heating the cases should ease bearing replacement.

    True False *Circle the answer.*

INSTRUCTOR VERIFICATION:

# Shop Assignment 11-2

Name ______________________ Date __________ Instructor ______________

## Two-Stroke Failure Modes and Symptoms

### Objective

After completing this activity, you should be able to demonstrate your knowledge of common two-stroke engine failure modes and their symptoms.

### Directions

Fill in the blank cells in the table below.

| Symptom | Remedy |
|---|---|
| Growling noise from engine while running. | |
| Transmission jumps out of one gear. | |
| Vehicle smokes excessively. Transmission oil low. | |
| Vehicle runs excessively lean. Jetting does not correct problem. | |
| Loud knocking noise when running. | |
| Crankshaft does not turn. | |
| Clutch does not disengage completely. | |

INSTRUCTOR VERIFICATION:

# Job Sheet 11-1

Name ______________________________ Date __________ Instructor ____________

## Two-Stroke Bottom-End Inspection

### Objective

After completing this job sheet, you should be able to correctly remove and replace the crankshaft from a modern two-stroke engine and check it for radial play.

### Directions

Using a training aid designated for this project, split the cases, remove the crankshaft and transmission shaft, check the crankshaft for radial play with V-blocks, and then replace the crankshaft and the transmission.

Note that most modern two-stroke engines are designed to be split from one side. Failure to split the cases from the correct side could dramatically increase the difficulty of this project. Check the service manual for directions. Also, because most crank bearings are a light press fit on the crankshaft, you may need to split the cases with a case splitter. This job sheet is written assuming you will need a case splitter. Some engines allow the case to slide off and back on without using a splitter.

Remember, when using a case splitter you cannot split the cases like a clam shell or severe damage could result. The case half must come off so that it always remains parallel to the opposite case half during the splitting process. Lightly tap on the transmission shafts with a dead blow hammer to keep the gap between the case halves even during removal.

### Tools and Equipment

Assorted hand tools, hydraulic press, case splitter, crank installation tool, dead blow hammer, V-blocks, and set of engine cases with no top-end installed

1. Double-check that all case screws or bolts are removed before attempting to split the cases.

2. Install the case splitter.

3. Turn the feed screw of the case splitter and observe the gap between the cases. It must remain even from front to back. Lightly tap on the transmission shafts with a dead blow hammer to keep the gap even.

4. Once the cases are split, remove the transmission shafts. Note the location of the thrust washers. Often they are of different thickness from one end to the other. Failure to replace the thrust washer on the correct shaft could result in a transmission shaft that does not turn when the cases are rejoined.

5. Remove the crankshaft from the opposite case. Try tapping it lightly with a dead blow hammer to see if it will slide out of its main bearing. If not, you will need to place the case half in a hydraulic press or use a crank removal tool to remove it from the case.

INSTRUCTOR VERIFICATION:

6. Support the ends of the crankshaft with V-blocks. Spin the shaft to make sure it was not knocked out of alignment during the removal process.

7. Use a crankshaft installation tool to pull the crankshaft through its main bearing.

8. Install the transmission shafts. You may need to mate the gears of both shafts together and insert the shafts simultaneously. Do not forget the thrust washers.

9. Slide the remaining case half over the transmission shafts and crank end.

10. Tap the case lightly with a dead blow hammer until the mating surfaces make contact.

11. Check to make sure the transmission shafts turn. If not, you will need to remove the case half again and check for the problem.

12. Make the engine ready for final inspection.

| INSTRUCTOR VERIFICATION: |
| --- |

CHAPTER 12

# Four-Stroke Engine Top-End Inspection

## Shop Assignment 12-1

Name ______________________ Date __________ Instructor ______________

## Four-Stroke Top-End Inspection Knowledge Assessment

### Objective

After completing this assessment, you should be able to demonstrate your knowledge of four-stroke engine top-end inspection procedures.

### Directions

1. Answer the following questions. Name two tests that should always be performed before a four-stroke engine is disassembled.
   a. ______________________
   b. ______________________

2. Technician A says that 15 percent leak down is too much and that the engine is showing signs of wear. Technician B says that 25 percent is the point at which an engine should be rebuilt. Who is right?

   ______________________________________________

3. What does air escaping from the exhaust pipe during a leak-down test tell the technician?

   ______________________________________________

4. Name a common tool for cleaning piston ring grooves.

   ______________________________________________

5. Can an old piston be used if it is still within specification?

   ______________________________________________

6. How do you know if the piston rings were installed correctly?

7. If the spark plug threads are ruined, must the head be replaced?

8. If the engine you are rebuilding has a warped head, what should be done?

9. How should valves be cleaned before inspecting them?

10. How should valve springs be installed in an engine?

11. Should valves be refaced?

12. How should new guides be prepared after installation?

13. Should the valve seat be cut if the valve guide was replaced?

14. Should titanium valves be lapped?

15. You are lapping in valves after the seats have been cut, and you notice that one of the valves is not hitting all the way around the valve face. What does this mean?

16. Why should a valve be tapped gently after the retainer and keepers are installed?

17. Where should the piston be when timing a camshaft on most engines?

18. What is the correct temperature for adjusting valves?

19. Why is it necessary to hone the cylinder wall when new piston rings are installed?

20. How can you tell whether the camshaft is correctly timed?

_______________________________________________

21. Name the three angles that most modern valves have.

a. ____________________

b. ____________________

c. ____________________

INSTRUCTOR VERIFICATION:

# Job Sheet 12-1

Name ______________________ Date __________ Instructor ______________

## Four-Stroke Top-End Inspection

### Objective

After completing this job sheet, you should be able to disassemble, inspect, and time the camshaft on a single-cylinder four-stroke engine.

### Directions

Using a training aid designated for this project, remove the camshaft, cylinder head, and cylinder and then reassemble the top end while correctly timing the camshaft.

### Tools and Equipment

Various hand tools

1. Obtain a training aid designated for this project. Use the service manual as your resource for step-by-step instructions.
2. Set the crankshaft to TDC. Remove the top cover.
3. Note the cam timing marks. Are they lined up properly?
   Yes No *Circle the answer.*
4. Remove the cam chain tensioner or adjust it to give you the maximum amount of chain slack.
5. Remove the camshaft.
6. Remove the cylinder.
7. Remove the piston.
8. Measure the piston diameter and the cylinder bore diameter. Calculate the piston to cylinder clearance: ________________ mm
9. What is the factory specification for piston-to-cylinder clearance? ________________ mm
10. Is the cylinder clearance within specification?
    Yes No *Circle the answer.*

11. Install the piston. Use the old wrist pin clips for this exercise if you are using a non-running engine. If this is a running engine, use new wrist pin clips.

12. Carefully install the piston into the cylinder. Make sure that the piston ring end gaps are properly staggered and that you do not snag any rings on the cylinder skirt. You may elect to use a piston ring compressor.

13. Install the cylinder head and any cam chain slippers or guides.

14. Install the camshaft. If you are using a Honda CRF230 for this project, make sure that the crankshaft is at TDC and that the "T" is adjacent to the index mark.

15. Install the cam chain tensioner. Make sure all of the slack is out of the cam chain.

16. Check your timing marks once the cam chain is tensioned.

17. Replace the valve cover and adjust the valves. Factory specification for valve lash:
    a. Intake: ______________________________ mm
    b. Exhaust: ______________________________ mm

18. Make the engine ready for final inspection.

INSTRUCTOR VERIFICATION:

# Job Sheet 12-2

Name ______________________________ Date __________ Instructor ______________

## Four-Stroke Cylinder Honing

### Objective

After completing this job sheet, you should be able to hone the cylinder on a typical four-stroke engine obtaining a 45-degree crosshatch finish.

### Directions

Using a cylinder designated for this project, clamp the cylinder securely in a vice and use a cylinder hone to produce a 45-degree crosshatch finish on the walls of the cylinder, being careful to keep the cylinder round.

### Tools and Equipment

Cylinder hone, drill motor

INSTRUCTOR VERIFICATION:

CHAPTER 13

# Four-Stroke Engine Bottom-End Inspection

## Shop Assignment 13-1

Name ______________________ Date __________ Instructor ______________

## Four-Stroke Bottom-End Inspection Knowledge Assessment

### Objective

After completing this assessment, you should be able to demonstrate your knowledge of four-stroke engine bottom-end inspection procedures.

### Directions

Answer the following questions.

1. What is used to measure the clearance of plain bearings?

   ______________________________________________

2. A motorcycle comes into the shop and the owner says that it is jumping out of gear. It is a vertically split engine. Does the top end need to be removed to service the transmission?

   Yes No *Circle the answer.*

3. A motorcycle comes into the shop and the owner says that it is jumping out of gear. It is a horizontally split engine. Does the top end need to be removed to service the transmission?

   Yes No *Circle the answer.*

4. What tools are needed to inspect an oil pump?

   ______________________________________________

5. The Babbitt main bearings of a CBR1000RR need replacement. What do you use to check the clearance between the crankshaft main journals and the bearings?

   ______________________________________________

   ______________________________________________

6. How are plain bearings identified?

_______________________________________________

7. Do the cases need to be split to service the clutch on a vertically split engine?

_______________________________________________

8. What tools are needed to check a crankshaft for run-out?

_______________________________________________

9. Can a multipiece crankshaft be rebuilt?

Yes No *Circle the answer.*

10. What is meant by a "cassette-type" transmission?
    a. A transmission that allows the gear sets to be changed for different tracks.
    b. A transmission with more than six speeds.
    c. A transmission that can be removed from the engine cases without splitting them.
    d. A transmission with a shift plate instead of a shift drum.

11. Name four things to look for when inspecting a clutch assembly.
    a. ______________________
    b. ______________________
    c. ______________________
    d. ______________________

12. How are new clutch plates prepared for installation?
    a. No preparation necessary
    b. Soak them in oil for a few minutes
    c. Use fine sandpaper to roughen the surface
    d. Clean them with contact cleaner

13. What is meant by a "live" seal?

_______________________________________________

14. How is Plastigage read?

_______________________________________________

15. Your best customer is an avid drag racer. He comes into the shop with a clutch that is completely burned up after a weekend of competition. While replacing the clutch on his bike, what else should be checked?

_______________________________________________

_______________________________________________

INSTRUCTOR VERIFICATION:

# Job Sheet 13-1

Name ______________________________ Date __________ Instructor ______________

## Four-Stroke Bottom-End Inspection

### Objective

After completing this job sheet, you should be able to disassemble, inspect, and correctly reassemble the bottom end of a four-stroke motorcycle.

### Directions

Using a training aid designated for this project, split the cases and remove the crankshaft. Check the crank for run-out and then reinstall it.

### Tools and Equipment

Vertically split crankcase, various hand tools, V-blocks, dial indicator

1. Obtain a training aid designated for this project. Use the service manual as your resource for step-by-step instructions.
2. Remove all case bolts and carefully split the cases.
3. Remove the crankshaft and set it on a crank jig or a pair of V-blocks.
4. Check the run-out of the crankshaft. Is it within specification?
   Yes No *Circle the answer.*
5. Are the main bearings smooth and quiet when you spin them?
   Yes No *Circle the answer.*
6. Is the small end of the connecting rod scored?
   Yes No *Circle the answer.*
7. Install the crankshaft.
8. If this training aid had a top end, reinstall it at this time and time the camshaft.
9. Make the engine ready for final inspection.

INSTRUCTOR VERIFICATION:

# Job Sheet 13-2

Name ______________________ Date __________ Instructor ______________

## Plain Bearing Inspection

### Objective

After completing this job sheet, you should be able to use Plastigage to measure plain bearing clearance.

### Directions

Using a training aid designated for this project, split the cases and insert Plastigage at every main bearing journal. Torque the cases to manufacturer's specification and then unbolt the cases and check the clearance.

### Tools and Equipment

Horizontally split cases, various hand tools, green Plastigage

1. Manufacturer's specification for main bearing clearance: ______________ mm
2. Number of main bearings in the engine: ______________
3. Color of Plastigage being used:
   Red Blue Green *Circle the answer.*
4. Case bolt torque: ________________ Nm
5. Measured clearance of bearings:
   a. Bearing #1: ____________ mm
   b. Bearing #2: ____________ mm
   c. Bearing #3: ____________ mm
   d. Bearing #4: ____________ mm
   e. Bearing #5: ____________ mm
6. Bearing codes on the crankshaft:
   a. __________ b. __________ c. __________ d. __________ e. __________
7. Bearing codes on the cases:
   a. __________ b. __________ c. __________ d. __________ e. __________

8. Consult the bearing code chart in the service manual and determine the correct color bearing for each journal.

   a. ____________ b. ____________ c. ____________ d. ____________ e. ____________

9. Thoroughly clean off the Plastigage with contact cleaner, and lightly oil the main bearings. Then reassemble the cases.

INSTRUCTOR VERIFICATION:

# Job Sheet 13-3

Name ________________________ Date __________ Instructor ______________

## Oil Pump Inspection

### Objective

After completing this job sheet, you should be able to accurately evaluate a trochiod oil pump.

### Directions

Using a training aid designated for this project, disassemble the oil pump and measure the tip clearance, rotor clearance, and overall condition of the oil pump.

### Tools and Equipment

Trochoid oil pump, hand tools, feeler gauges, straightedge

1. Disassemble the oil pump and note any scoring on the sides of the rotor or tips.

   Good Bad *Circle the answer.*

2. Manufacturer's specification for rotor-to-body clearance: ______________ mm

3. Measured clearance between the rotor and the body: ______________ mm

4. Manufacturer's specification for tip clearance: ______________ mm

5. Measured clearance between the rotor and the body: ______________ mm

6. Using a straightedge set along the top of the pump body, measure the clearance between the top of the body and the top of the rotor.

7. Manufacturer's specification for rotor-to-body clearance: ______________ mm

8. Measured rotor-to-body clearance: ______________ mm

9. Reassemble the oil pump and make it ready for final inspection.

INSTRUCTOR VERIFICATION:

# CHAPTER 14 Electrical Fundamentals

## Shop Assignment 14-1

Name ______________________________ Date __________ Instructor ____________

## Electrical Fundamentals Knowledge Assessment

### Objective

After completing this assessment, you should be able to demonstrate your knowledge of electrical fundamentals.

### Directions

Answer the following questions.

1. An electrical component that has an anode and a cathode is called a ________________
2. Should an ammeter be hooked up in **series** or **parallel** when measuring amperage? *Circle the answer.*
3. To measure battery voltage on a motorcycle, set the meter to ________________ volts.
4. Where is AC voltage found on a motorcycle?
   ________________________________________________
5. Calculate the resistance in a 12-volt circuit that has 2 amps of current: ________________
6. What would be an acceptable voltage drop across an ignition switch? ____________ volts DC.
7. How much voltage should there be after the load? ________________
8. Motorcycle manufacturers generally use the **conventional** or **electron** theory. *Circle the answer.*

9. A transistor is an example of a ______________________________

10. Name a type of semiconductor found on modern motorcycles: ______________

11. ______________________________ is a measure of electrical pressure.

12. ______________________________ is a measure of current flow.

13. ______________________________ is a measurement of opposition to current flow.

14. What is the unit of measurement for resistance? ______________________

15. Is house current **AC** or **DC**? *Circle the answer.*

16. The entire schematic of a modern motorcycle can cover several pages. What is the portion that deals only with the circuit you are interested in called?

    ______________________________________________________________

17. Are color codes consistent among all motorcycle manufacturers?

    Yes No *Circle the answer.*

18. A customer comes into the shop and complains that his battery goes dead. Which circuit is probably at fault?

    Ignition Charging Starting Lighting *Circle the answer.*

19. An unwanted path to ground before the load is called a ______________________

20. What is the meter setting for measuring charging system current?

    ______________________________________________________________

21. What is the meter setting for measuring voltage drop across a switch?

    ______________________________________________________________

22. What is the meter setting for measuring charging voltage at the battery?

    ______________________________________________________________

23. What would happen if a meter set to DC volts is used to measure charging current?

    ______________________________________________________________

24. You are measuring the resistance of the stator coil in a properly operating charging system. The manual says it should have 0.2 ohms of resistance, but the reading you get is 110 ohms. What did you forget to do? ______________________________

25. A circuit with more than one path to ground is called a ______________________

| INSTRUCTOR VERIFICATION: |
| --- |

# Shop Assignment 14-2

Name ______________________________ Date __________ Instructor ______________

## Reading Schematic Diagrams

### Objective

After completing this activity, you should be able to read a simple schematic and create a block diagram.

### Directions

Answer the following questions using the schematic on page 136.

A

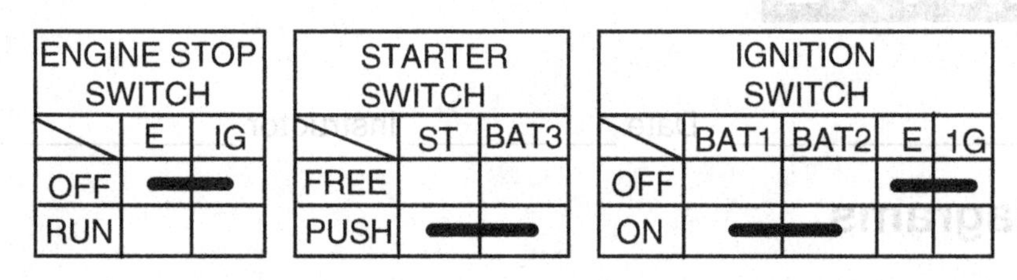

| ENGINE STOP SWITCH | E | IG |
|---|---|---|
| OFF | — | — |
| RUN | | |

| STARTER SWITCH | ST | BAT3 |
|---|---|---|
| FREE | | |
| PUSH | — | — |

| IGNITION SWITCH | BAT1 | BAT2 | E | 1G |
|---|---|---|---|---|
| OFF | | | — | — |
| ON | — | — | | |

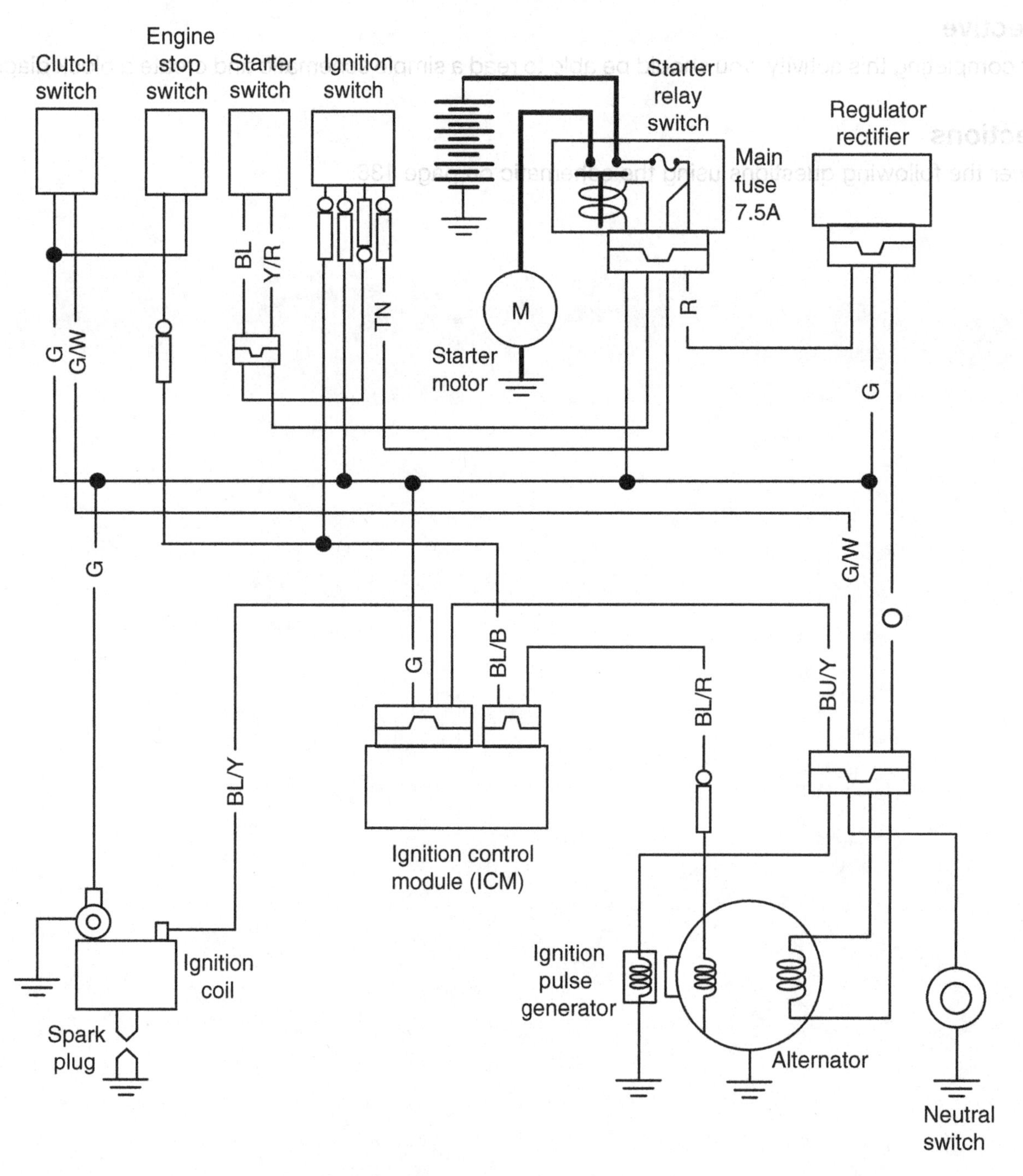


1. What color is the ground wire in the schematic? ______________________________

2. Does this bike have to be in neutral to start? ______________________________

   Yes                No                *Circle the answer.*

3. Does the clutch switch have to be closed for this bike to start?

   Yes No *Circle the answer.*

4. What color wire supplies voltage to the start switch? ______________________

5. Where does the start switch get its power? ______________________

6. Which wires have continuity (are connected to each other) when the ignition switch is in the off position? ______________________

7. Using the schematic picture on the previous page, make a block diagram of the starting system. Your diagram should include the battery, starter motor, starter relay, neutral switch, ignition switch, starter switch, and clutch switch. Use a separate piece of paper to practice making your block diagram and then use the box provided below for your final version.

| INSTRUCTOR VERIFICATION: |
| --- |

# Job Sheet 14-1

Name ______________________________ Date __________ Instructor ______________

## Meter Usage

### Objective

After completing this job sheet, you should be able to use your meter to measure volts, amperage, and Ohms on a motorcycle.

### Directions

Use your multimeter, the training aid designated for this project, and the appropriate service manual to answer the following questions.

### Tools and Equipment

Multimeter

1. Locate the motorcycle's battery and place the black multimeter lead on the negative battery post. Place the red meter lead on the positive post. Set the meter to DCV. What is the reading? ____________________ DCV

2. Was a **series** or **parallel** connection made? *Circle the answer.*

3. Remove the headlight and pull off the connector at the back. Consult the service manual and decide which of the three terminals on the connector is for the low beam. Plug the red meter lead into that terminal. Place the black meter lead on any bare metal part of the vehicle. Set your meter to DCV. Turn on the ignition switch and set the headlight switch to low beam. What is the reading? ____________________ DCV

4. Consult the service manual and determine which of the three terminals is the ground. Place the black meter lead in this terminal and the red meter lead in the low beam terminal. What is the reading? ____________________ DCV

5. Disconnect the negative battery cable. Set your meter to DCA. Move the meter leads to the correct jacks on the meter for measuring amperage. Hook your meter's black lead to the negative battery post. Hook your red meter lead to the negative battery cable. Turn on the ignition switch. Do NOT start the vehicle! What is the reading? ____________________ DCV

6. Set your meter to ohms and move the meter leads to the proper jacks for measuring resistance. Place one meter lead on one terminal of the headlight you removed in step 3. Place the other meter lead on the ground terminal. What is the reading? ____________________ ohms

| INSTRUCTOR VERIFICATION: |
| --- |

# Job Sheet 14-2

Name ______________________________ Date __________ Instructor ______________

## Voltage Drop Testing

### Objective

After completing this job sheet, you should be able to do simple voltage drop testing.

### Directions

Using a training aid designated for this project, follow the instructions and answer the following questions.

1. Set your meter to DCV. Move the meter leads to the correct meter jacks for measuring DC voltage. Place the red meter lead on one side of the main fuse. (Do NOT remove the fuse.) Place the other meter lead on the opposite side of the main fuse. Turn on the ignition switch. What is the reading? ____________________DCV

2. You just did a voltage drop test. You measured how much voltage was dropped across the main fuse. Find a connector in the wiring harness. The connector must be part of an active circuit when the ignition switch is turned on. Place your meter leads across the connector and turn on the ignition switch. What is the reading? ____________________DCV

INSTRUCTOR VERIFICATION:

CHAPTER 15

# Motorcycle Charging Systems

## Shop Assignment 15-1

Name ______________________ Date __________ Instructor __________

## Charging Systems Knowledge Assessment

### Objective

After completing this assessment, you should be able to demonstrate your knowledge of charging system fundamentals.

### Directions

Answer the following questions.

1. All charging systems function in basically the same way.

   True False *Circle the answer.*

2. Technician A says that an automotive-style charger works fine for charging motorcycle batteries as long as it is set to a low charge level. Technician B says that a good motorcycle-style charger should be a constant current type. Who is correct?

   A B *Circle the answer.*

3. A half-wave charging system uses a single diode to rectify the current.

   True False *Circle the answer.*

4. How is a three-phase excited-field charging system regulated?
   a. By grounding one or more stator leads
   b. By looping excess current back to the stator coil
   c. By reducing the current flowing through the field coil
   d. By shunting excess current to ground

5. A battery can be recharged by either AC or DC current, depending on the type of battery.

   True    False    *Circle the answer.*

6. A customer complains that his battery goes dead while sitting. Which components can be eliminated as faulty? *Circle all that apply.*

   a. Stator coil
   b. Rotor
   c. Rectifier
   d. Wiring

7. You start the motorcycle and connect an ammeter to the battery. The meter reads +2 DCA. Is this system charging?

   Yes    No    *Circle the answer.*

8. A customer complains that her battery goes dead when her bike sits for a day. You connect your ammeter to the battery and note that with the key off, the meter reads 2 DCA. Is this normal current drain?

   Yes    No    *Circle the answer.*

9. Technician A says that you must use a known good battery to check a charging system. Technician B says that you should use the battery that came with the vehicle. Who is correct?

   A    B    *Circle the answer.*

10. How do you check for a bad field coil?

    a. Do a resistance check and compare your readings to the factory specification.
    b. Field coils cannot go bad because they are solid state.
    c. Measure the current draw across the slip rings.
    d. Check the AC output and compare to the factory specification.

11. To check a stator coil for AC output, you unplug it from the charging system and connect the meter leads to each side of the stator. What is the next step?

    a. Start the engine and observe the AC output.
    b. Set the meter to ohms and check its resistance.
    c. Measure the voltage drop across the coil.
    d. Never check an unplugged stator because it will burn out.

12. What would be a typical resistance for each pair of legs of a stator coil?

    a. 1–10 ohms
    b. 2–100 ohms
    c. 0.2–1 ohms
    d. Infinity

13. What would be the typical charge voltage when checked at the battery for a properly operating charging system?
    a. 8–12 DDC
    b. 10–20 VAC
    c. 13–15 VDC
    d. 1–2 ohms

14. How is a three-phase permanent magnet system regulated?
    a. By reducing current to the field coil
    b. By shunting one or more of the stator leads to ground
    c. By disconnecting one or more stator leads from the rectifier
    d. By grounding the rectifier leads

15. A technician wants to test a three-phase excited-field charging system, so he connects the meter in series to the battery and sets the meter to amps. He tries to crank the engine over, but it will not turn over. What did he do wrong?
    a. He started the motorcycle with the meter in the circuit.
    b. He should have put the meter in parallel for an amp check.
    c. He did not let the machine warm up.
    d. He forgot to turn on all the lights.

16. List three advantages of the CAN bus technology.
    a. ______________________________
    b. ______________________________
    c. ______________________________

17. How will CAN bus technology help you as a technician diagnose problems with motorcycle electrical systems?

    ______________________________________________________________

    ______________________________________________________________

    ______________________________________________________________

| INSTRUCTOR VERIFICATION: |
| --- |

# Job Sheet 15-1

Name ______________________________ Date __________ Instructor ______________

## Charging System Inspection #1

### Objective

After completing this job sheet, you should be able to accurately evaluate the charging system on any modern motorcycle.

### Directions

Check at least one example of each type of charging system available. Since some tests listed on this job sheet do not apply to all charging systems, mark "NA" on tests that do not apply. Use a multimeter, a training aid designated for this project, and the appropriate service manual to answer the following questions.

### Tools and Equipment

Multimeter

1. Identify the type of charging system by the number of leads on the stator.
   a. 1; half wave
   b. 2; full wave
   c. 3; three phase
2. Is this a **permanent magnet** or an **excited-field** type of charging system? *Circle the answer.*
3. **Key-off amp draw test.** Determine if the electrical system is draining the battery when the key is switched off by hooking the meter in series to the negative side of the battery. Observe any current draw with the key off. Note that some bikes with clocks or computer memory should draw 1–2 milliamps with the key off.

   Current draw: ______________________________
4. **Charging voltage test.** Check the charging voltage at the battery. Hook the meter in parallel to the battery and let the vehicle idle.

   Voltage at idle: ______________________________
5. **Charging current test.** Remove the main fuse and hook the meter in series at the fuse holder after configuring the meter to test DCA. You should now be able to start the machine with the meter hooked up in series without damage to it. Observe the charging current at idle.

   Current at idle: ______________________________

6. **Stator output test.** Unplug all the stator leads and hook the meter to one pair of leads with the meter set at ACV. Test all three pairs of leads. Start the machine and record the voltage at idle.

   ACV at idle: 1–2 ______________ ACV

   1–3 ______________ ACV

   2–3 ______________ ACV

7. **Field coil test.** If the motorcycle you are testing has an excited field, you may be able to check the field coil and brushes. Refer to the service manual for details on testing the field coil.
   a. What is the manual specification for brush length? _________ mm
   b. What is the manual specification for field coil resistance? ________ ohms

8. **Rectifier test.** Unplug the regulator/rectifier and check the forward and reverse bias of the diodes in the rectifier. Note that there should be two checks for testing a half-wave system, four checks for a full-wave system, and six checks for a three-phase system.
   a. Forward bias: ______________

   ______________

   ______________

   b. Reverse bias: ______________

   ______________

   ______________

9. Is the charging system working correctly?

   Yes No *Circle the answer.*

10. If you circled "No," explain what is wrong with the system.

    ______________________________________________

    ______________________________________________

    ______________________________________________

INSTRUCTOR VERIFICATION:

# Job Sheet 15-2

Name ______________________ Date __________ Instructor ______________

## Charging System Inspection #2

### Objective

After completing this job sheet, you should be able to accurately evaluate the charging system on any modern motorcycle.

### Directions

Check at least one example of each type of charging system available. Since some tests listed on this job sheet do not apply to all charging systems, mark "NA" on tests that do not apply. Use a multimeter, a training aid designated for this project, and the appropriate service manual to answer the following questions.

### Tools and Equipment

Multimeter

1. Identify the type of charging system by the number of leads on the stator.
   a. 1; half wave
   b. 2; full wave
   c. 3; three phase
2. Is this a **permanent magnet** or an **excited-field** type of charging system? *Circle the answer*.
3. **Key-off amp draw test.** Determine if the electrical system is draining the battery when the key is switched off by hooking the meter in series to the negative side of the battery. Observe any current draw with the key off. Note that some bikes with clocks or computer memory should draw 1–2 milliamps with the key off.

   Current draw: ______________________________
4. **Charging voltage test.** Check the charging voltage at the battery. Hook the meter in parallel to the battery and let the vehicle idle.

   Voltage at idle: ______________________________
5. **Charging current test.** Remove the main fuse and hook the meter in series at the fuse holder after configuring the meter to test DCA. You should now be able to start the machine with the meter hooked up in series without damage to it. Observe the charging current at idle.

   Current at idle: ______________________________

6. **Stator output test.** Unplug all the stator leads and hook the meter to one pair of leads with the meter set at ACV. Test all three pairs of leads. Start the machine and record the voltage at idle.

   ACV at idle: 1–2 ________________ ACV

   1–3 ________________ ACV

   2–3 ________________ ACV

7. **Field coil test.** If the motorcycle you are testing has an excited field, you may be able to check the field coil and brushes. Refer to the service manual for details on testing the field coil.

   a. What is the manual specification for brush length? ________________ mm

   b. What is the manual specification for field coil resistance? ________________ ohms

8. **Rectifier test.** Unplug the regulator/rectifier and check the forward and reverse bias of the diodes in the rectifier. Note that there should be two checks for testing a half-wave system, four checks for a full-wave system, and six checks for a three-phase system.

   a. Forward bias: ________________

   ________________

   ________________

   b. Reverse bias: ________________

   ________________

   ________________

9. Is the charging system working correctly?

   Yes No *Circle the answer.*

10. If you circled "No," explain what is wrong with the system.

    ________________________________________________

    ________________________________________________

    ________________________________________________

INSTRUCTOR VERIFICATION:

# Job Sheet 15-3

Name ______________________________ Date __________ Instructor ______________

## Charging System Inspection #3

### Objective

After completing this job sheet, you should be able to accurately evaluate the charging system on any modern motorcycle.

### Directions

Check at least one example of each type of charging system available. Since some tests listed on this job sheet do not apply to all charging systems, mark "NA" on tests that do not apply. Use a multimeter, a training aid designated for this project, and the appropriate service manual to answer the following questions.

### Tools and Equipment

Multimeter

1. Identify the type of charging system by the number of leads on the stator.
   a. 1; half wave
   b. 2; full wave
   c. 3; three phase

2. Is this a **permanent magnet** or an **excited-field** type of charging system? *Circle the answer.*

3. **Key-off amp draw test.** Determine if the electrical system is draining the battery when the key is switched off by hooking the meter in series to the negative side of the battery. Observe any current draw with the key off. Note that some bikes with clocks or computer memory should draw 1–2 milliamps with the key off.
   Current draw: ______________________________

4. **Charging voltage test.** Check the charging voltage at the battery. Hook the meter in parallel to the battery and let the vehicle idle.
   Voltage at idle: ______________________________

5. **Charging current test.** Remove the main fuse and hook the meter in series at the fuse holder after configuring the meter to test DCA. You should now be able to start the machine with the meter hooked up in series without damage to it. Observe the charging current at idle.
   Current at idle: ______________________________

6. **Stator output test.** Unplug all the stator leads and hook the meter to one pair of leads with the meter set at ACV. Test all three pairs of leads. Start the machine and record the voltage at idle.

   ACV at idle: 1–2 ________________ ACV

   1–3 ________________ ACV

   2–3 ________________ ACV

7. **Field coil test.** If the motorcycle you are testing has an excited field, you may be able to check the field coil and brushes. Refer to the manual for details on testing the field coil.

   a. What is the manual specification for brush length? ________________ mm

   b. What is the manual specification for field coil resistance? ________________ ohms

8. **Rectifier test.** Unplug the regulator/rectifier and check the forward and reverse bias of the diodes in the rectifier. Note that there should be two checks for testing a half-wave system, four checks for a full-wave system, and six checks for a three-phase system.

   a. Forward bias: ________________

   ________________

   ________________

   b. Reverse bias: ________________

   ________________

   ________________

9. Is the charging system working correctly?

   Yes No *Circle the answer.*

10. If you circled "No," explain what is wrong with the system.

    ________________________________________

    ________________________________________

    ________________________________________

INSTRUCTOR VERIFICATION:

# Job Sheet 15-4

Name ______________________________ Date __________ Instructor ______________

## Charging System Inspection #4

### Objective

After completing this job sheet, you should be able to accurately evaluate the charging system on any modern motorcycle.

### Directions

Check at least one example of each type of charging system available. Since some tests listed on this job sheet do not apply to all charging systems, mark "NA" on tests that do not apply. Use a multimeter, a training aid designated for this project, and the appropriate service manual to answer the following questions.

### Tools and Equipment

Multimeter

1. Identify the type of charging system by the number of leads on the stator.
   a. 1; half wave
   b. 2; full wave
   c. 3; three phase

2. Is this a **permanent magnet** or an **excited-field** type of charging system? *Circle the answer*.

3. **Key-off amp draw test.** Determine if the electrical system is draining the battery when the key is switched off by hooking the meter in series to the negative side of the battery. Observe any current draw with the key off. Note that some bikes with clocks or computer memory should draw 1–2 milliamps with the key off.
   Current draw: ______________________________

4. **Charging voltage test**. Check the charging voltage at the battery. Hook the meter in parallel to the battery and let the vehicle idle.
   Voltage at idle: ______________________________

5. **Charging current test.** Remove the main fuse and hook the meter in series at the fuse holder after configuring the meter to test DCA. You should now be able to start the machine with the meter hooked up in series without damage to it. Observe the charging current at idle.
   Current at idle: ______________________________

6. **Stator output test.** Unplug all the stator leads and hook the meter to one pair of leads with the meter set at ACV. Test all three pairs of leads. Start the machine and record the voltage at idle.

   ACV at idle: 1–2 _______________ ACV

   1–3 _______________ ACV

   2–3 _______________ ACV

7. **Field coil test.** If the motorcycle you are testing has an excited field, you may be able to check the field coil and brushes. Refer to the service manual for details on testing the field coil.

   a. What is the manual specification for brush length? _______________ mm

   b. What is the manual specification for field coil resistance? _______________ ohms

8. **Rectifier test.** Unplug the regulator/rectifier and check the forward and reverse bias of the diodes in the rectifier. Note that there should be two checks for testing a half-wave system, four checks for a full-wave system, and six checks for a three-phase system.

   a. Forward bias: _______________

   _______________

   _______________

   b. Reverse bias: _______________

   _______________

   _______________

9. Is the charging system working correctly?

   Yes No *Circle the answer.*

10. If you circled "No," explain what is wrong with the system.

    _______________________________________________

    _______________________________________________

    _______________________________________________

INSTRUCTOR VERIFICATION:

CHAPTER 16

# Ignition and Electric Starting Systems

## Shop Assignment 16-1

### Ignition Systems Knowledge Assessment

Name ______________________________ Date __________ Instructor ____________

### Objective

After completing this assessment, you should be able to demonstrate your knowledge of charging system fundamentals.

### Directions

Answer the following questions.

1. Choose the three functions an ignition system performs.
   a. Provide a hot spark, prevent the spark plug tip from burning, and deliver a properly timed spark.
   b. Maintain a spark long enough to ignite the fuel mixture, deliver a spark to each cylinder at the correct time, and provide a hot spark.
   c. Maintain a spark long enough to ignite the fuel mixture, deliver a spark well in advance of top dead center (TDC), and provide a spark that has ample reserve energy.

2. What type of ignition system uses an AC power source?
   a. Points-type ignition system
   b. Capacitor discharge ignition system
   c. Battery powered ignition system
   d. Digital ignition system

3. What type of ignition system fires the spark plug every 360 degrees of crankshaft rotation?
   a. Wasted spark system
   b. Multiple discharge system
   c. Digital system
   d. Hall-effect system

4. A Hall-effect sensor is used to ______________________________ the spark.
   a. Time
   b. Increase
   c. Decrease
   d. Delay

5. Moto-cross bikes generally have a(n) ______________________________ ignition system.
   a. Battery powered
   b. Low-tension magneto
   c. High-tension magneto
   d. AC powered

6. An ignition coil is essentially a ______________________.
   a. Flux capacitor
   b. Transformer
   c. Zener diode
   d. Little battery

7. A coil with more secondary windings than primary windings should ______________________.
   a. Step up the primary voltage
   b. Step down the primary voltage
   c. Step up the primary current
   d. Provide earlier ignition timing

8. An ignition coil sparks the spark plug when current to the primary side of the coil is ____________.
   a. Switched off and on
   b. Energized by the pulser coil
   c. Grounded to the frame
   d. Induced into the stator windings

9. A spark plug with a long heat path is ______________________ than a plug with a short heat path.
   a. Hotter
   b. Colder
   c. Bigger
   d. Smaller

10. Spark plugs have resistors built into them to ________________.
    a. Make them last longer
    b. Increase secondary voltage
    c. Reduce radio frequency interference
    d. Reduce preignition

11. Precious metal spark plugs are used to ________________.
    a. Reduce the voltage needed to spark them and to last longer
    b. Increase the spark volume
    c. Keep the plug running cooler
    d. Keep the plug clean

12. Breaker point ignition systems are no longer used because they ________________.
    a. Are too complicated
    b. Require frequent maintenance
    c. Operate at too high a temperature
    d. Are too slow to deliver enough sparks to the plug

13. When a set of points is arcing it means the ________________.
    a. Coil is bad
    b. Points are not timed properly
    c. Condenser is bad
    d. Spark plug resistor is bad

14. The ignition switch on an AC-powered ignition system ________________.
    a. Grounds the system to stop the engine
    b. Cuts the voltage to the primary side of the coil to stop the engine

15. A starter motor can draw over 120 amps of current while cranking the engine over.
    True False *Circle the answer.*

| INSTRUCTOR VERIFICATION: |
| --- |

# Job Sheet 16-1

Name ______________________________ Date __________ Instructor ____________

## AC CDI Ignition System Inspection

### Objective

After completing this job sheet, you should be able to accurately evaluate the ignition system on an AC CDI ignition system.

### Directions

Use a multimeter, a training aid designated for this project, and the appropriate service manual to answer the following questions.

### Tools and Equipment

Multimeter

1. Remove the spark plug and insert it into the plug cap. Lay the plug on the cylinder head and crank/kick the engine with all switches in the run position. Did you see a spark?

   Yes No *Circle the answer.*

2. Does the ignition and/or stop switch ground the system or cut battery voltage to the primary side of the ignition coil?
   a. Grounds the system
   b. Cuts current to the coil

### Exciter Coil Check

3. Disconnect the leads from the source or exciter coil and attach the meter leads to them. Set the meter to ACV and turn over the engine briskly. Note that the leads are usually found at the CDI module.
   a. Record the voltage: ______________________ ACV
   b. Factory specification for exciter coil voltage: ______________________ ACV

4. Set the meter to read ohms and measure the resistance of the exciter coil.
   a. Record the resistance: ______________________ ohms
   b. Factory specification for exciter coil resistance: ______________________ ohms

5. Reconnect the exciter coil leads.

## Trigger Coil Check

6. Disconnect the leads from the pulser/trigger coil. Briskly turn over the engine. Note that the leads are usually found at the CDI module.
   a. Record the voltage: ______________________________ ACV
   b. Factory specification for pulser coil voltage: ______________________________ ACV

7. Set the meter to read ohms and measure the resistance of the trigger coil.
   a. Record the resistance: ______________________________ ohms
   b. Factory specification for trigger coil resistance: ______________________________ ohms

8. Reconnect the trigger coil leads.

## Stop Switch Check

9. Disconnect the leads from the stop switch. Measure the resistance of the stop switch in both the on and off positions.
   a. On: ______________________________ ohms
   b. Off: ______________________________ ohms

10. Is the stop switch wired in **series** or **parallel** on this vehicle? *Circle the answer.*

11. Reconnect all leads and check to see if there is a spark at the plug as you did in step 1. Did you see a spark?

    Yes No *Circle the answer.*

12. Did you find any components that did not meet factory specifications?

    Yes No *Circle the answer.*

13. Make the vehicle ready for final inspection.

| INSTRUCTOR VERIFICATION: |
| --- |

# Job Sheet 16-2

Name ______________________________ Date __________ Instructor ______________

## DC CDI Ignition System Inspection

### Objective

After completing this job sheet, you should be able to accurately evaluate the ignition system on a DC CDI ignition system.

### Directions

Use a multimeter, a training aid designated for this project, and the appropriate service manual to answer the following questions.

### Tools and Equipment

Multimeter

1. Remove the spark plug and insert it into the plug cap. Lay the plug on the cylinder head and crank/kick the engine with all switches in the run position. Did you see a spark?

   Yes No *Circle the answer.*

2. Does the ignition and/or stop switch ground the system or cut battery voltage to the primary side of the ignition coil?

   a. Grounds the system

   b. Cuts current to the coil

### Battery Voltage Check

3. The DC CDI ignition system differs from an AC CDI system in that the battery supplies the source voltage. Consult the schematic in the service manual to see where the battery voltage is connected to the ignition module. Set the meter to DCV and check whether 12 volts are present with all switches in the run position.

   Record the voltage: ______________________ DCV

### Trigger Coil Check

4. Disconnect the leads from the pulser/trigger coil. Briskly turn over the engine. Note that the leads are usually found at the CDI module.

   a. Record the voltage: ______________________ ACV

   b. Factory specification for pulser coil voltage: ______________________ ACV

5. Set the meter to read ohms and measure the resistance of the trigger coil.
   a. Record the resistance: ______________________ ohms
   b. Factory specification for trigger coil resistance: ______________________ ohms
6. Reconnect the trigger coil leads.

## Stop Switch Check

7. Disconnect the leads from the stop switch. Measure the resistance of the stop switch in both the on and off positions.
   a. On: ______________________ ohms
   b. Off: ______________________ ohms
8. Is the stop switch wired in **series** or **parallel** on this vehicle? *Circle the answer.*
9. Reconnect all leads and check to see if there is a spark at the plug as you did in step 1. Did you see a spark?

   Yes No *Circle the answer.*
10. Did you find any components that did not meet factory specifications?

    Yes No *Circle the answer.*
11. Make the vehicle ready for final inspection.

INSTRUCTOR VERIFICATION:

# Job Sheet 16-3

Name ______________________________ Date __________ Instructor ______________

## Pointless Electronic Ignition System Inspection

### Objective

After completing this job sheet, you should be able to accurately evaluate the ignition system on an electronic pointless ignition system.

### Directions

Use a multimeter, a training aid designated for this project, and the appropriate service manual to answer the following questions.

### Tools and Equipment

Multimeter

1. Remove the spark plug and insert it into the plug cap. Lay the plug on the cylinder head and crank/kick the engine with all switches in the run position. Did you see a spark?

   Yes No *Circle the answer.*

2. Does the ignition and/or stop switch ground the system or cut battery voltage to the primary side of the ignition coil?

   a. Grounds the system

   b. Cuts current to the coil

### Battery Voltage Check

3. Remove the leads from the primary side of the ignition coil and check to see whether 12 volts are present with all switches in the run position.

   Record the voltage: ________________________ DCV

4. Consult the service manual schematic to find what color leads supply 12 DCV to the ignition module. Remove those leads and check for battery voltage.

   Record the voltage: ________________________ DCV

### Trigger Coil Check

5. Disconnect the leads from the pulser/trigger coil. Briskly turn over the engine. Note that the leads are usually found at the ignition module.

   a. Record the voltage: ________________________ ACV

   b. Factory specification for pulser coil voltage: ________________________ ACV

6. Set the meter to read ohms and measure the resistance of the trigger coil.
   a. Record the resistance: ______________________ ohms
   b. Factory specification for trigger coil resistance: ______________________ ohms

7. Reconnect the trigger coil leads.

## Stop Switch Check

8. Disconnect the leads from the stop switch. Measure the resistance of the stop switch in both the on and off positions.
   a. On: ______________________ ohms
   b. Off: ______________________ ohms

9. Is the stop switch wired in **series** or **parallel** on this vehicle? *Circle the answer.*

10. Reconnect all leads and check to see if there is a spark at the plug as you did in step 1. Did you see a spark?

    Yes No *Circle the answer.*

11. Did you find any components that did not meet factory specifications?

    Yes No *Circle the answer.*

12. Make the vehicle ready for final inspection.

INSTRUCTOR VERIFICATION:

# Frames and Suspension

## Shop Assignment 17-1

Name ______________________ Date __________ Instructor __________

## Frame and Suspension Knowledge Assessment

### Objective

After completing this assessment, you should be able to demonstrate your knowledge of frames and suspension systems.

### Directions

Answer the following questions.

1. Match the term with its definition.

| | | |
|---|---|---|
| a. Swing arm | _____ | dynamic damping control |
| b. Rising rate | _____ | a type of frame in which the engine is a stressed member |
| c. Progressive wound | _____ | the length of the vehicle from axle to axle |
| d. Rake | _____ | a shock that uses nitrogen to keep oil from foaming |
| e. Trail | _____ | steering caster |
| f. Wheelbase | _____ | a frame that wraps around the engine |
| g. Pivotless | _____ | the upward stroke of the suspension |
| h. Perimeter | _____ | a type of frame popular on scooters |
| i. De Carbon | _____ | the downward action of the suspension |
| j. Underbone | _____ | a frame made from short sections of tubing |
| k. Trellis | _____ | the most widely used type of front suspension |
| l. Diamond | _____ | front suspension that eliminates braking drive |
| m. Telelever | _____ | distance from line through front axle and steering stem |
| n. Backbone | _____ | wide beam from which engine is suspended |

o. Rebound stroke ______ where the swing arm pivots in the engine cases
p. Compression stroke ______ suspension that gets stiffer as it compresses
q. Pneumatic spring ______ a spring that changes the distance from each coil
r. Telescopic ______ mounting bracket for rear wheel assembly
s. DDC ______ type of spring that is replacing steel fork springs

2. Adding more rake and trail
   a. Increases stability
   b. Decreases stability
   c. Increases the wheelbase
   d. Both (a) and (c)

3. Modern dampers have much more rebound damping than compression damping.
   True False *Circle the answer.*

4. A smaller damper rod orifice creates more ______________________________ than a larger one.
   a. Friction
   b. Damping
   c. Trail
   d. Fade

5. A shock absorber must have an air space to accommodate the
   a. Damper oil
   b. Nitrogen
   c. Damper rod
   d. Shim stack

6. All single-shock motorcycles use ______________________________ rear suspension.
   a. Falling-rate
   b. Rising-rate
   c. Single-rate
   d. Progressively wound

7. A shock spring with tighter wound coils at the top than the bottom is called a ______________________________ rate spring.
   a. Dual
   b. Single
   c. Progressive
   d. Inverted

8. Cartridge forks often have adjusters to allow for changing the

a. Rake and trail
b. Compression and rebound damping
c. Hop and wobble
d. Ride height and race sag

9. Single shock suspension is superior to twin shock suspension.
   True False *Circle the answer.*

10. The de Carbon shock uses ______________________ gas in its bladder.
    a. Helium
    b. Oxygen
    c. Nitrogen
    d. Hydrogen

11. A separate function fork has a ______________________ in only one of the fork tubes.

12. A balance chamber is found on ______________________ forks.

INSTRUCTOR VERIFICATION:

# Shop Assignment 17-2

Name ______________________________ Date __________ Instructor ______________

## Identifying Frame Types

### Objective

After completing this activity sheet, you should be able to identify all of the common frame types.

### Directions

Answer the following question.

Identify the frames pictured below on the blank line to the right of each photo.

A

______________________________

B

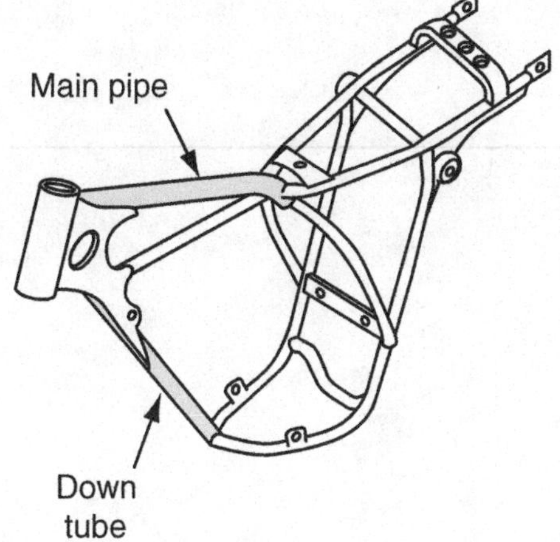


______________________________

C

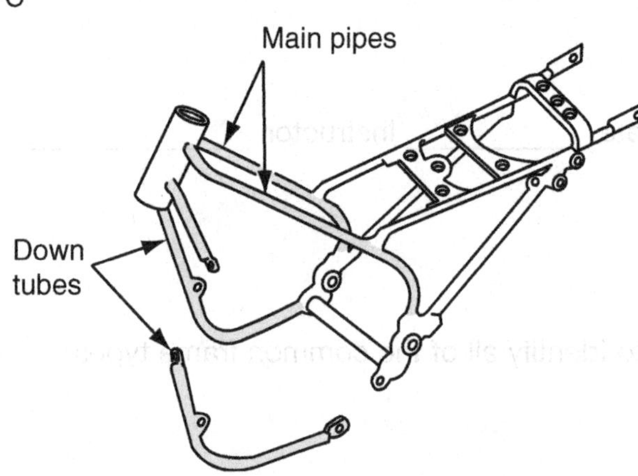
Main pipes
Down tubes

______________________________

D

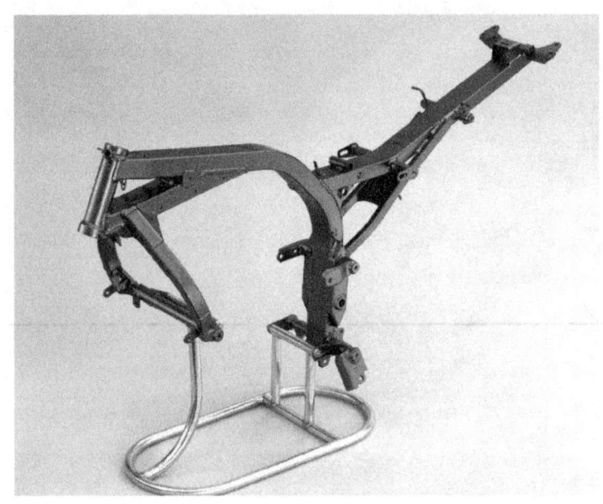

______________________________

E

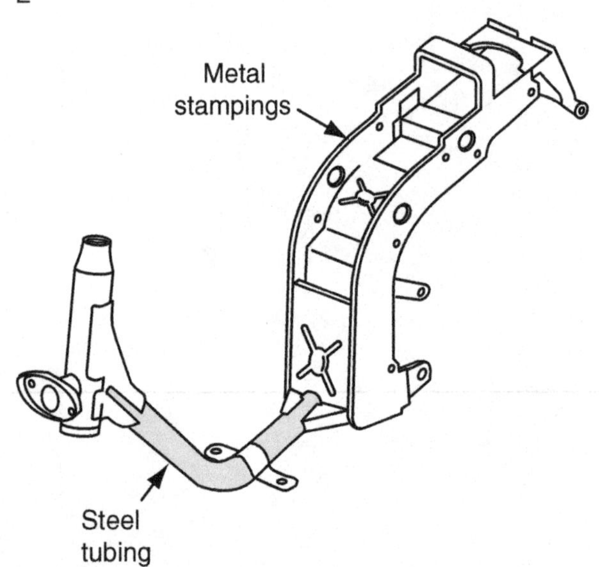
Metal stampings
Steel tubing

______________________________

F

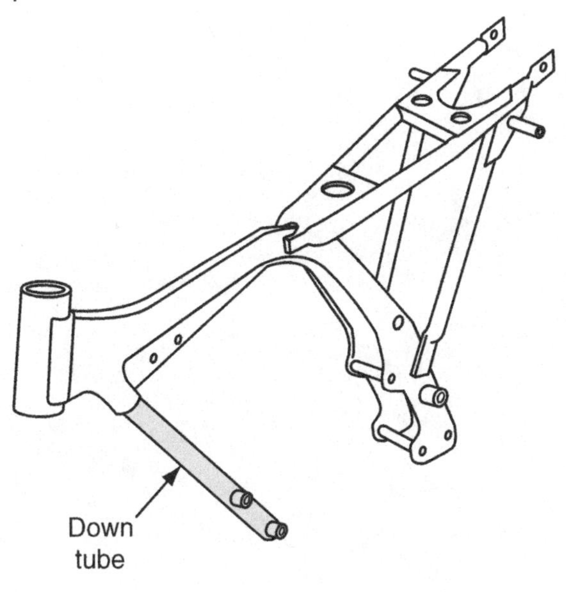

G

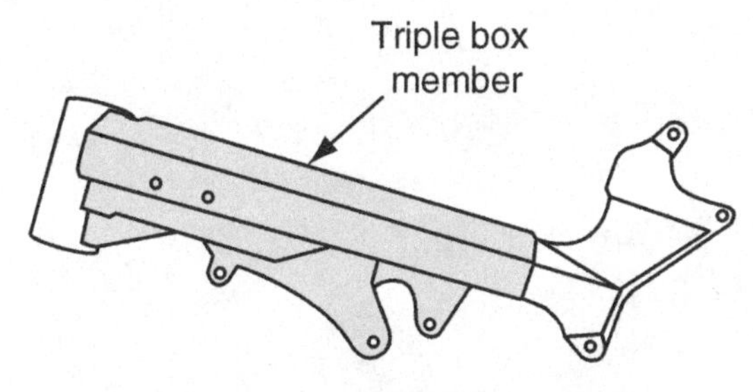


H

| INSTRUCTOR VERIFICATION: |
| --- |

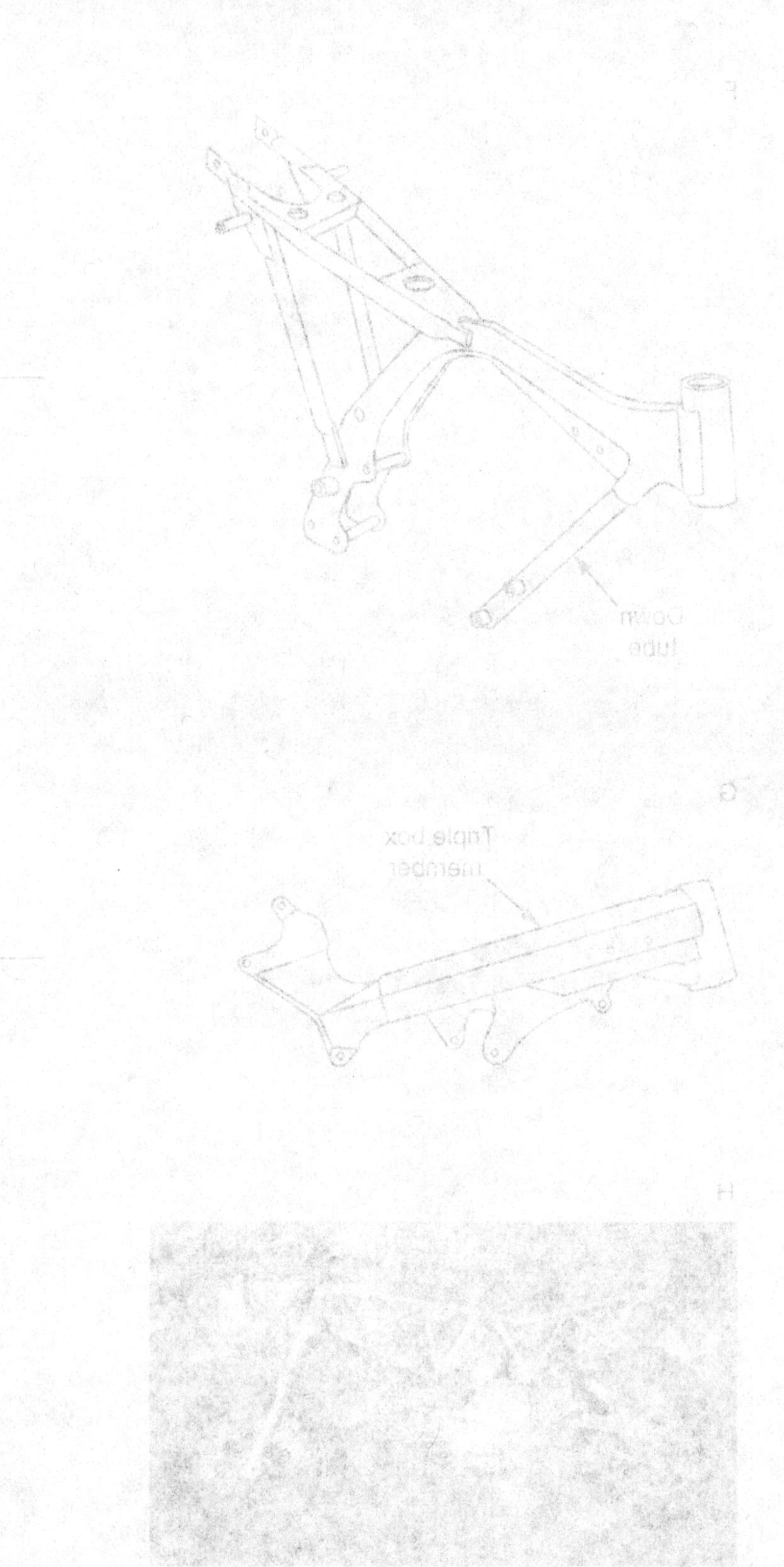

# Shop Assignment 17-3

Name ______________________________ Date __________ Instructor ______________

## Identifying Suspension Components

### Objective

After completing this activity sheet, you should be able to identify commonly used suspension components.

### Directions

Answer the following question.

Identify the suspension components pictured below on the blank line to the right of each photo.

I

______________________________

J

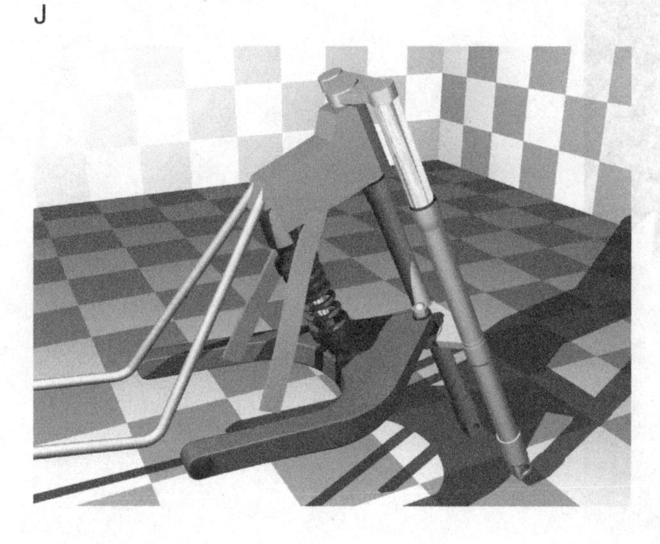

______________________________

K

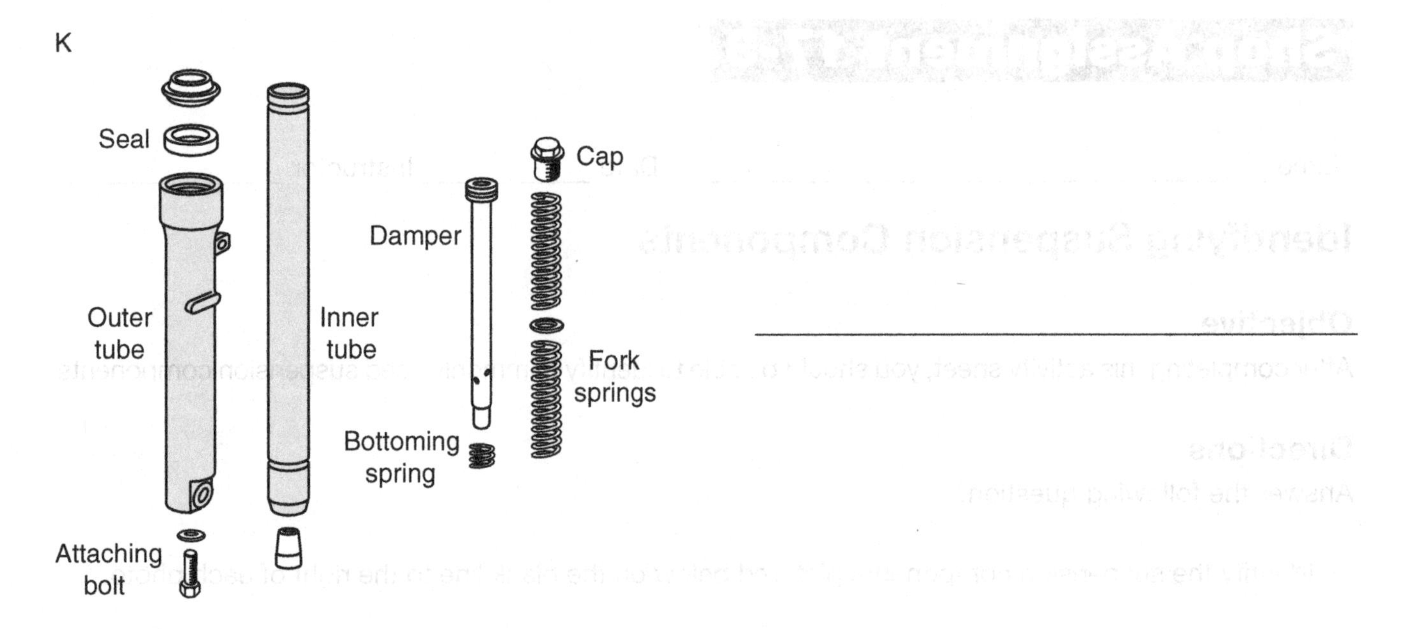
Seal
Cap
Damper
Outer
tube
Inner
tube
Fork
springs
Bottoming
spring
Attaching
bolt

L

NISSIN

M

N

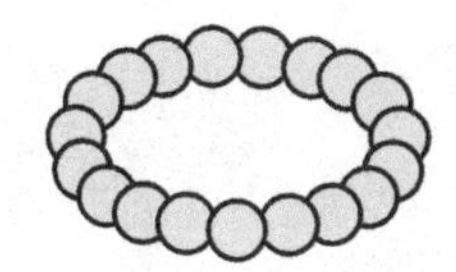

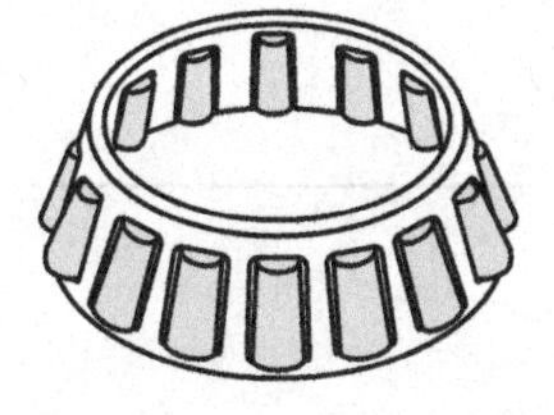

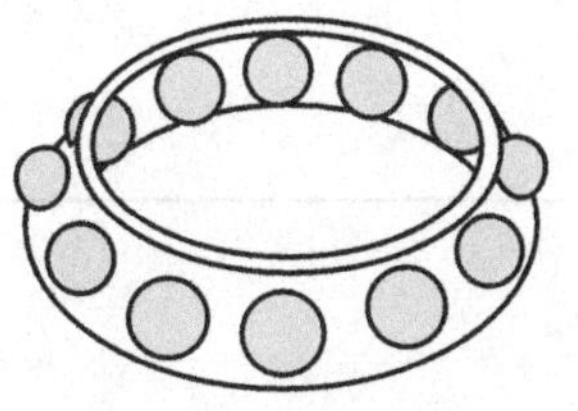

O

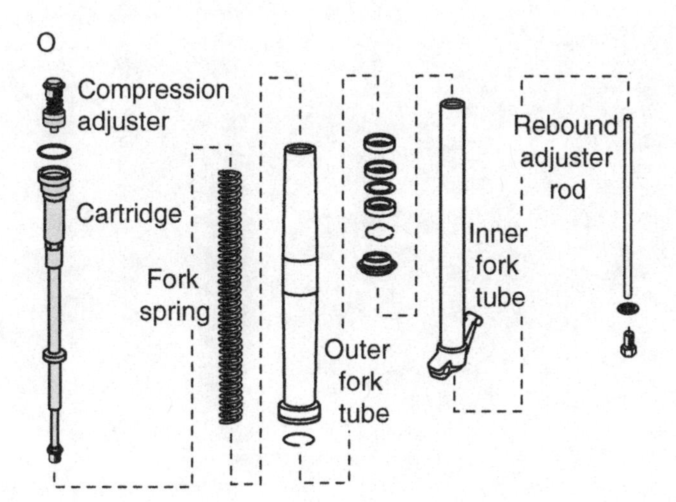
Compression adjuster
Cartridge
Fork spring
Outer fork tube
Inner fork tube
Rebound adjuster rod

P

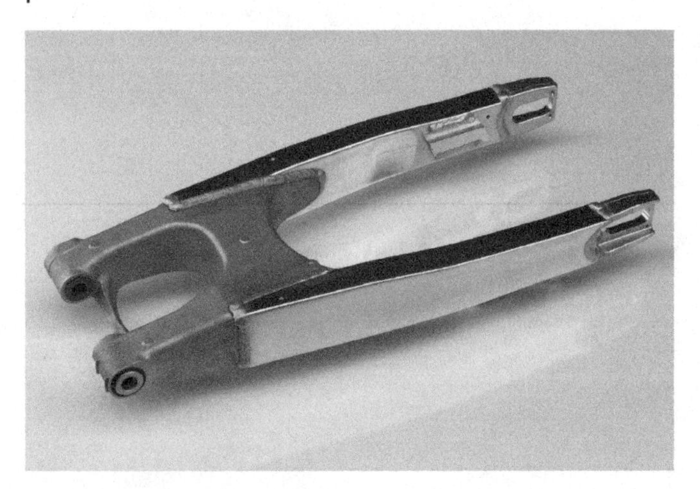

Q

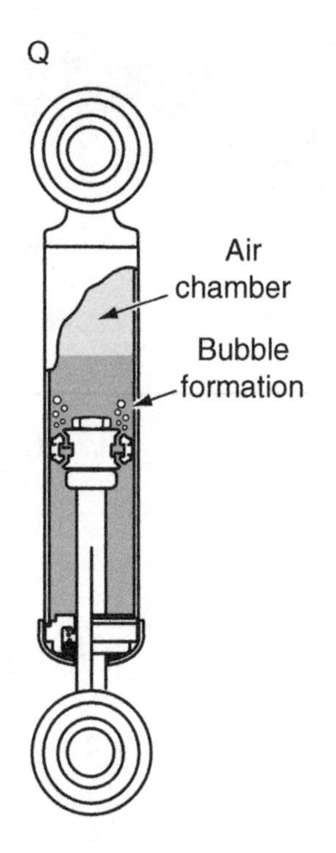


INSTRUCTOR VERIFICATION:

# Job Sheet 17-1

Name ______________________ Date __________ Instructor ____________

## Setting Shock Spring Preload

### Objective

The most basic suspension adjustment is setting the shock spring preload. After completing this job sheet, you should be able to set the rear shock spring preload correctly for a given weight rider.

### Directions

Find an assistant and set the shock spring preload for his or her weight.

### Tools and Equipment

Preload ring spanner wrench, tape measure

1. Preload is adjusted so that one-third of the total travel is used up when the rider is sitting in the saddle. Use the service manual to find total suspension travel and divide that number by 3.

   Example: Total rear suspension travel = 300 mm
   Suspension sag target = 300/3 = **100 mm**

2. Suspend the motorcycle so that the rear wheel is off the ground. Pick a spot near the rear fender and measure from there to the rear axle.

   Record this measurement: ______________________ mm

3. Put both wheels on the ground and have your rider/assistant sit on the machine and measure the same two spots as accurately as possible.

   Record the second measurement: ______________________ mm

4. Subtract the second number from the first. The result is the suspension sag. Add or remove spring preload to arrive at the target number calculated in step 1.

If you have more sag than your target, tighten the spring preload until you arrive at the target sag number. If the machine reads less than your target sag number, loosen the preload collar.

On some machines, you may have as few as three different choices for spring preload. If this is the case for the motorcycle you are working on, just get the measurement as close as possible. On modern dirt bikes, you should be able to hit the target exactly.

| INSTRUCTOR VERIFICATION: |
| --- |

# Job Sheet 17-2

Name ______________________________ Date __________ Instructor ______________

## Servicing Steering Head Bearings

### Objective

After completing this job sheet, you should be able to properly inspect, evaluate, and service steering head bearings according to manufacturer specifications.

### Directions

Using a training aid designated for this project and the appropriate service manual, complete the steps detailed below.

### Tools and Equipment

Various hand tools, steering stem spanner wrench

1. Support the vehicle so that the front wheel is off the ground. Swing the front wheel back and forth, noting any notchiness or unusual stiffness as the wheel moves from side to side.

2. Consult the factory service manual and remove the steering components necessary to expose the steering stem nut.

3. Carefully remove the stem nut, and the steering stem can be removed from the frame. Note that if you are servicing a steering stem supported by loose ball bearings, you should be ready to catch the bearings that fall out of their races.

4. Clean and inspect the bearings. Note the number of ball bearings that you have and compare that number to the number specified in the service manual.

5. Clean and inspect the races in the steering head and on the bottom of the steering stem. If you note any pitting of the hard chrome surface or brinneling where the race is dimpled by the bearings, the races must be replaced.

6. Coat the races with heavy wheel bearing grease and carefully place the bearings in their races.

7. Reassemble the steering stem in the steering head following the directions in the service manual.

8. Put the initial torque on the steering stem nut and swing the stem back and forth a few times. If the stem does not move back and forth smoothly, disassemble and start over again.

9. Reassemble the remaining front-end components and make the vehicle ready for final inspection.

| INSTRUCTOR VERIFICATION: |
| --- |

# Job Sheet 17-3

Name ______________________________ Date __________ Instructor ____________

## Adjusting Steering Head Bearings

### Objective

After completing this job sheet you, should be able to properly adjust steering head bearings according to manufacturer specifications.

### Directions

Using a training aid designated for this project and the appropriate service manual, complete the steps detailed below.

### Tools and Equipment

Various hand tools, torque wrench, spring (fish) scale, string, steering stem spanner wrench

1. Raise the front end of the vehicle off the ground so that the front forks are free to pivot.
2. Loosen the steering head bearing preload collar.
3. Torque the preload collar to factory specification.
4. Tie a string around one of the fork tubes. Attach the spring scale to the string. Pull on the scale until the fork deflects. Write down the number of pounds needed to deflect the fork.
   ____________________ lbs.
5. Your target is 2–3 lbs. of force to deflect the front end. Adjust the torque on the steering bearing preload collar until it takes 2–3 lbs. to deflect the front end.
6. Make the motorcycle ready for final inspection.

INSTRUCTOR VERIFICATION:

# Brakes, Wheels, and Tires

## Shop Assignment 18-1

Name ______________________ Date __________ Instructor ______________

## Frame and Suspension Knowledge Assessment

### Objective

After completing this assessment, you should be able to demonstrate your knowledge of the brakes, wheels, and tires used on modern motorcycles.

### Directions

Answer the following questions.

1. A single leading shoe drum brake has one brake shoe.

   True  False  *Circle the answer.*

2. To accurately check a brake rotor for warpage you need a ______________.
   a. Dial caliper
   b. Degree wheel
   c. Dial indicator
   d. Feeler gauge

3. A customer comes into your shop and complains that his hydraulic front brake is spongy. You check the brake pads, hoses, and fluid level, which all appear to be well within the manufacturer's specification. What should be done next?
   a. Add more brake fluid
   b. Replace the brake rotor
   c. Change to DOT 4 brake fluid
   d. Bleed the brakes

4. This code appears on the sidewall of the tire you must replace: 170/60 R 18 73 H. What type of tire is indicated?
   a. Radial
   b. Bias ply
   c. Belted bias ply

5. What is the maximum allowable sustained speed for the tire specified in Question 4?
   a. 100 mph
   b. 100 kmh
   c. 210 kmh
   d. 240 kmh

6. How wide is the tire specified in Question 4?
   a. 18 mm
   b. 60 mm
   c. 73 mm
   d. 170 mm

7. Technician A says that ABS brakes can stop a motorcycle quicker than a human operating the brakes manually. Technician B says that ABS brakes are designed to stop a rider safely when traction is compromised by rain or sand on the road. Who is correct?

   A B *Circle the answer.*

8. You are measuring a brake rotor because you suspect that it is worn beyond its limits. Where is the minimum allowable thickness for a brake rotor found?
   a. In the owner's manual
   b. On the brake rotor

9. A customer comes into your shop with a nearly new rear tire but it has a puncture. To avoid buying a new tire she requests that you repair the tubeless tire by inserting a tube. Is this acceptable?

   True False *Circle the answer.*

10. The free play of a rear drum brake can be adjusted at
    a. The brake pedal
    b. The adjuster
    c. The swing arm
    d. Drum brakes do not need adjustment

11. A linked braking system can actuate both front and rear brakes from the brake pedal.

    True False *Circle the answer.*

12. Only bikes with hydraulic brakes have linked brakes.

    True        False        *Circle the answer.*

13. What protects the tube from abrasion by the spoke nipples on a dirt bike wheel?
    a. Nothing, nipples never touch the tube
    b. Nipple caps
    c. A rim band
    d. Duct tape

14. ABSs use _______________ brake fluid.
    a. Dot 3
    b. Dot 4
    c. Dot 5
    d. Dot 5.1

15. A customer comes into your shop and requests a rear tire much wider than the original one on his motorcycle. What do you tell him?
    a. "A larger tire will rub the swing arm and cause a blow out."
    b. "The tire you want may cause the bike to weave and wobble at high speeds."
    c. "I'll get right on it."
    d. Both (a) and (b)

| INSTRUCTOR VERIFICATION: |
| --- |

# Job Sheet 18-1

Name ______________________________ Date __________ Instructor ______________

## Master Cylinder Inspection

### Objective

After completing this job sheet, you should be able to correctly disassemble, inspect, and reassemble a master cylinder assembly.

### Directions

Using a training aid designated for this project and the factory service manual, remove the master cylinder, inspect its plunger and seals, and reassemble it. If you are going to work on Job Sheet 18-3, do not put brake fluid in the system until 18-3 is completed.

### Tools and Equipment

Brake fluid, brake bleeder, 90 internal snap-ring pliers, assorted hand tools

1. Drain the brake fluid from the system.
2. Undo the brake hose and remove the master cylinder.
3. Using the service manual as your guide, remove the snap ring that holds the plunger in the master cylinder. Note the position of the return spring. It must be returned to its original position and orientation.
4. Closely inspect the seals and cups on the plunger. Look for cuts or nicks.
5. Did you find any imperfections in the seals or cups?

   Yes No *Circle the answer.*
6. Reinstall the return spring and plunger assembly.
7. Reinstall the master cylinder and brake hose with new crush washers but do not yet add the brake fluid. If you are going to work on Job Sheet 18-3, start it now. If not, proceed to Step 9.
8. Replace the brake fluid and proceed to Job Sheet 18-4, which will instruct you to bleed the system.
9. Make the vehicle ready for inspection.

INSTRUCTOR VERIFICATION:

# Job Sheet 18-2

Name ______________________ Date __________ Instructor ______________

## Brake Caliper Inspection

### Objective

After completing this job sheet, you should be able to correctly disassemble, inspect, and reassemble a caliper assembly.

### Directions

Using a training aid designated for this project and the factory service manual, remove the caliper, inspect its piston and seals, and reassemble it.

### Tools and Equipment

Brake piston pliers, brake fluid, brake bleeder, assorted hand tools, regulated compressed air supply

1. Using the service manual as your guide, remove the caliper assembly. If the vehicle designated for this project has two calipers, remove only one.

2. Remove the piston(s) by pulling it out with brake piston pliers. The pliers are inserted into the open end of the piston and have teeth on the outside of the jaws. This will allow you to grip the piston and pull it free.

   If brake piston pliers are not available, you may use compressed air to blow the piston out. **Caution**: Use a regulated air supply. Using more than a few pounds of air pressure could force the piston out at a high speed and cause injury.

   If the caliper has more than one piston, blow out the first one, then reinsert it part way and hold it in place while blowing out the second one.

3. Inspect the piston for scoring and rust damage. Look carefully at the seals. They must be free of nicks and tears. Did you find any imperfections in the seals?

   Yes No *Circle the answer.*

4. Reassemble the caliper, remount it, and attach the brake hose with new crush washers. Attach the caliper to the vehicle.

INSTRUCTOR VERIFICATION:

5. Proceed to Job Sheet 18-4.

INSTRUCTOR VERIFICATION:

# Job Sheet 18-3

Name ______________________ Date __________ Instructor ______________

## Brake Bleeding

### Objective

After completing this job sheet, you should be able to correctly bleed a brake system.

### Directions

Using a training aid designated for this project and the factory service manual, bleed the brake system.

### Tools and Equipment

Brake bleeder, 8 mm wrench

1. Fill the master cylinder with brake fluid.
2. Attach the brake bleeder to the bleeder valve and open it with an 8 mm wrench.
3. Pump the bleeder so that fluid is pulled through the system and starts to fill the bleeder cup.
4. Close the bleeder valve and check to see whether the brake lever is firm. Chances are that the system still has some air in it. If this is the case, slowly stroke the brake lever several times. Hold the lever in and open the bleeder valve. You may see air bubbles in the bleeder line. Close the valve and repeat this procedure until no more bubbles appear in the bleeder line while making sure that the master cylinder never runs dry. If it does, you must begin again at step 3. Repeat the procedure until the brake lever feels firm.
5. Top off the master cylinder, replace the cap, and prepare the vehicle for final inspection.

INSTRUCTOR VERIFICATION:

# Job Sheet 18-4

Name ______________________ Date __________ Instructor ______________

## Tube-Type Tire Changing

### Objective

After completing this job sheet, you should be able to correctly change out an inner tube in a tube-type tire.

### Directions

Using a training aid designated for this project and the factory service manual, remove the front wheel, remove the inner tube, and replace it with the correct amount of air pressure.

### Tools and Equipment

Bead breaker, two tire irons, valve core tool

1. Remove the valve core.

2. Remove the valve stem nut.

3. Use a bead breaker to loosen the bead all the way around on both sides of the tire.

4. Lever the bead off the rim using small "bites" while being careful not to snag the inner tube. If this happens, the tube will be torn and will not hold air. Pull the inner tube out of the tire. Now is a good time to check the rim band. It should be in good condition. If not, replace it.

5. Stuff the inner tube back into the tire making sure the valve core fits into the hole in the rim provided for it. Make sure the tube is evenly distributed in the tire and not bunched up.

6. Replace the valve core and add just enough air for the tube to take its normal shape.

7. This is the most critical part of the operation. Carefully lever the bead back onto the tire being especially careful not to pinch the tube between the rim and the tire iron (adding enough air for the tube to take shape helps avoid this). If the tire is very hard to lever back on the rim, try letting out some of the air.

8. Inflate the tire until it seats properly and adjust the air pressure to the amount designated by the instructor. This should usually be about 20 psi.

9. If the valve stem sits at an angle in its hole, let the air out, break the bead again, and spin the tire until the valve stem sticks out straight. Replace the valve stem nut.

10. Refill the tire with the correct amount of air.

11. Mount the wheel on the vehicle. Spin the wheel to make sure the tire is beaded all the way around the wheel.

12. Remember to pump the brake back up if the motorcycle is equipped with a disc brake.

| INSTRUCTOR VERIFICATION: |
| --- |

# Job Sheet 18-5

Name ______________________________ Date __________ Instructor ______________

## Tire Balancing

### Objective

After completing this job sheet, you should be able to accurately balance a tire.

### Directions

Using a training aid designated for this project, mount a wheel and tire assembly onto a tire balancer. Refer to the detailed instructions for using the tire balancer.

### Tools and Equipment

Tire balancer, wheel weights

1. Mount the tire on the tire balancer.
2. Remove any old wheel weights.
3. Spin the tire by hand or engage the automatic spin cycle of the tire balancer, whichever is appropriate for the balancer being used.
4. Place the recommended amount of wheel weight in the spot designated by the balancer.
5. Spin the tire once more to check your work.
6. Prepare the wheel for final inspection.

| INSTRUCTOR VERIFICATION: |
| --- |

# Job Sheet 18-6

Name ______________________ Date __________ Instructor ______________

## Spoke Wheel Lacing

### Objective

After completing this job sheet, you should be able to identify inner and outer spokes, find the correct holes in the rim for the spokes, and properly lace the wheel.

### Directions

Using a 36-spoke rim designated for this project and the factory service manual, remove all spoke nipples and spokes and then re-lace the wheel.

### Tools and Equipment

Screwdriver, spoke wrench, truing stand, masking tape

1. Identify the inner spokes. They have more than a 90-degree bend at the end. The length of the inner spokes runs along the inside of the hub flange.

2. Identify the outer spokes. They have less than a 90-degree bend at the end. The length of the outer spokes runs along the outside of the hub flange.

3. Unscrew the nipples from the rim and remove all spokes. Group them into two piles, inner and outer.

4. Install the inner spokes in the hub first. Start by putting the first spoke in any spoke hole of the hub and then attach a piece of tape to it.

5. Install the second inner spoke up through the lower flange. Mark this spoke with a piece of tape.

6. Install eight more inner spokes on top, skipping every other hole in the hub. Make these spokes point in a counterclockwise direction.

7. Install eight more inner spokes up through the lower hub flange. Make these spokes point in a clockwise direction.

8. Lay the rim over the hub and put the upper spoke marked with tape in the first hole clockwise from the valve stem hole. Keep it in position by screwing a nipple on it.

9. Lace the remaining inner upper spokes to every fourth spoke hole in the rim and hold them in position with the nipples. Twist the hub in a clockwise direction until the spokes are pulled tight.

10. Carefully turn the wheel over and lace the lower taped spoke into the seventh spoke hole counterclockwise from the first spoke. This spoke should point in the opposite direction of the installed spokes.

11. Lace the rest of the inner spokes to every fourth hole.

12. Install nine outer spokes in the lower flange and connect them to the rim so that they point in a counterclockwise direction.

13. Install nine outer spokes in the top flange and connect them to the remaining holes.

14. Tighten all of the nipples so that only two threads show above each nipple.

| INSTRUCTOR VERIFICATION: |
| --- |

# Job Sheet 18-7

Name ______________________________ Date __________ Instructor ______________

## Spoke Rim Truing

### Objective

After completing this job sheet, you should be able to correctly true a spoke rim to 0.5 mm for hop and wobble.

### Directions

Using a training aid designated for this project and the factory service manual, true the rim using a dial indicator to accurately measure for hop and wobble.

### Tools and Equipment

Truing stand, spoke wrench, dial indicator, dial indicator stand

1. Mount the rim from Job Sheet 18-6 in a truing stand. Tighten all of the spokes so that only two threads show. The rim should be fairly true but loose.

2. Tighten all spokes another one-half turn, then repeat until all are evenly snug. Use your fingers to tighten and NOT a screwdriver.

3. Mount the dial indicator so that it measures the radial movement or "hop" of the rim. Always correct the hop before the wobble.

4. Find the high point in the hop with the dial indicator and tighten the spokes a little with a screwdriver. Then find the low point in the hop and loosen the spokes if they are tight.

5. Continue with this procedure until the hop is within 0.5 mm. Then work on the wobble.

6. Mount the dial indicator so that it measures the lateral movement or "wobble" of the rim.

7. Find the low point in the wobble with the dial indicator. As the rim moves away from the dial indicator, tighten the spokes to pull the rim toward the indicator.

8. Find the high point in the wobble and loosen those spokes a little if they are very tight.

9. Continue with this procedure until the dial indicator shows about 0.5 mm of wobble. Strive to get all of the spokes evenly tight. Remember the biggest mistake in rim truing is overtightening the spokes.

10. Prepare the rim for final inspection.

INSTRUCTOR VERIFICATION:

CHAPTER 19

# Maintenance and Emission Controls

## Shop Assignment 19-1

Name ______________________ Date __________ Instructor __________

## Maintenance and Emission Controls Knowledge Assessment

### Objective

After completing this assessment, you should be able to demonstrate your knowledge of motorcycle maintenance procedures and emission control operation.

### Directions

Answer the following questions.

1. What are the three types of emission control systems found on street bikes?
   a. Crankcase, evaporative, exhaust
   b. Evaporative, recirculation, crankcase
   c. Exhaust, intake, catalytic
   d. Carbon dioxide, carbon monoxide, nitrogen oxide

2. The purge control valve is part of which emission system?
   a. Exhaust
   b. Evaporative
   c. Crankcase
   d. Recirculation

3. A carburetor clean is usually part of a tune-up.
   True False *Circle the answer.*

4. The specific gravity of a(n) ______________________ battery should be checked during a tune-up.
   a. MF
   b. AGM
   c. Conventional
   d. Nicad

5. A customer comes into the shop because his bike tipped over and now refuses to start. You find that the charcoal canister is full of raw fuel, which is fouling the spark plugs. The customer requests that you remove it. What should be done?
   a. Charge the customer and remove it.
   b. Explain that it is against the law to remove it.
   c. Open it up, dump out the raw fuel, and reinstall it.
   d. None of the above.

6. The reed valves in the air injection system prevent:
   a. Carbon from forming in the exhaust ports
   b. Backfiring on deceleration
   c. Exhaust gasses from backing up into the airbox
   d. A tapping noise at idle

7. What is the best source for the maintenance schedule for a motorcycle?
   a. Manufacturer's Web site
   b. Owner's manual
   c. Service manual
   d. Both (b) and (c)

8. It is not necessary to replace the oil filter with every oil change if the oil change interval is within the maintenance schedule.

   True      False      *Circle the answer.*

9. A centrifugal oil filter:
   a. Filters oil through a set of wire mesh screens
   b. Has no filtering elements
   c. Is mainly found on small-displacement motorcycle engines
   d. Both (b) and (c)

10. Checking the oil level on a bike equipped with a sight glass is done:
    a. Only when the machine is upright
    b. When the machine is upright or on the side stand
    c. Only when the machine is on the side stand
    d. Before the engine is started

11. A compression test cannot be done on a two-stroke engine.

    True False *Circle the answer.*

12. Checking the spark plugs during a tune-up can tell a great deal about the engine's condition.

    True False *Circle the answer.*

13. The paper air filter on a high mileage machine should be ______________________ during the course of a major tune-up.
    a. Blown out with compressed air
    b. Washed
    c. Replaced
    d. None of the above

14. All of the carburetors on a multicylinder motorcycle should be synchronized to:
    a. The center carburetor
    b. Each other
    c. The base carb
    d. 100 mm Hg

15. Check tire wear by inspecting:
    a. The TWI
    b. For cupping
    c. For cords showing
    d. All of the above

INSTRUCTOR VERIFICATION:

# Job Sheet 19-1

Name ______________________ Date __________ Instructor ______________

## Four-Stroke 12,000-Mile Service

### Objective

After completing this job sheet, you should be able to correctly service a motorcycle according to the maintenance schedule listed in Table 19-2 in the textbook and be able to perform a leakdown and compression test.

### Directions

Using a training aid designated for this project and the factory service manual, perform a 12,000-mile service.

### Tools and Equipment

Various hand tools, compression gauge, leakdown tester

1. Consult Table 19-2 on page 429 of the textbook. Inspect the items listed in the maintenance schedule. If you have any questions concerning a service procedure, consult the instructor.

2. **Compression Test:**

   Manufacturer's specification: ______________________

   Measured compression: Cyl 1 ______________________

   Cyl 2 ______________________

   Cyl 3 ______________________

   Cyl 4 ______________________

3. **Leakdown Test:**

   Measured leakdown: Cyl 1 ______________________

   Cyl 2 ______________________

   Cyl 3 ______________________

   Cyl 4 ______________________

4. **Valve Clearance Inspection:**

   Manufacturer's specification: Intake ______________________

   Exhaust ______________________

   Measured valve lash: Intake Cyl 1 __________/__________

   Cyl 2 __________/__________

   Cyl 3 __________/__________

   Cyl 4 __________/__________

Measured valve lash: Exhaust Cyl 1 ______________/______________

Cyl 2 ______________/______________

Cyl 3 ______________/______________

Cyl 4 ______________/______________

5. Inspect, but do not replace, the items listed below. Check the box after you have completed each inspection.
   - a. Fuel line ☐
   - b. Throttle operation ☐
   - c. Choke operation ☐
   - d. Air cleaner ☐
   - e. Spark plugs ☐
   - f. Engine oil ☐
   - g. Oil filter ☐
   - h. Carburetor synchronize ☐
   - i. Idle speed ☐
   - j. Coolant ☐
   - k. Secondary air ☐
   - l. Evaporative emission ☐
   - m. Brake pads ☐
   - n. Brake light ☐
   - o. Headlight ☐
   - p. Clutch cable ☐
   - q. Side stand ☐
   - r. Fasteners ☐
   - s. Tires ☐
   - t. Steering bearings ☐
6. Make the vehicle ready for final inspection.
7. Note any additional items that need service.

   ______________________________________________________________

INSTRUCTOR VERIFICATION:

# Job Sheet 19-2

Name ______________________________ Date __________ Instructor ______________

## Troubleshoot the Crankcase Emission System

### Objective

After completing this job sheet, you should be able to correctly troubleshoot the crankcase emission system.

### Directions

Using a training aid designated for this project and the factory service manual, troubleshoot the crankcase emission system.

### Tools and Equipment

Various hand tools

1. Customer's contention: Bike smokes excessively during acceleration but runs well during slow speed operation.
2. Write down what was found during your inspection.

 ______________________________________________________________

 ______________________________________________________________

3. Explain how to correct the problem.

 ______________________________________________________________

INSTRUCTOR VERIFICATION:

# Job Sheet 19-3

Name ______________________ Date __________ Instructor ______________

## Troubleshoot the Evaporative Emission System

### Objective

After completing this job sheet, you should be able to correctly troubleshoot the evaporative emission system.

### Directions

Using a training aid designated for this project and the factory service manual, inspect the evaporative emission system.

### Tools and Equipment

Various hand tools, vacuum pump

1. Customer's contention: Bike started and died; will not take throttle.
2. Write down what was found during your inspection.

   ______________________________________________

   ______________________________________________

3. Explain how to correct the problem.

   ______________________________________________

INSTRUCTOR VERIFICATION:

# Job Sheet 19-4

Name ______________________________ Date __________ Instructor ______________

## Troubleshoot the Air Injection Emission System

### Objective

After completing this job sheet, you should be able to correctly troubleshoot the air injection emission system.

### Directions

Using a training aid designated for this project and the factory service manual, inspect the air injection emission system.

### Tools and Equipment

Various hand tools, vacuum pump

1. Customer's contention: Bike backfires during deceleration.
2. Write down what was found during your inspection.
   ______________________________________________________________
   ______________________________________________________________
3. Explain how to correct the problem.
   ______________________________________________________________

INSTRUCTOR VERIFICATION:

CHAPTER 20

# Motorcycle Troubleshooting

## Shop Assignment 20-1

Name ______________________ Date __________ Instructor __________

## Troubleshooting Knowledge Assessment

### Objective

After completing this assessment, you should be able to demonstrate your knowledge of the principles of effective troubleshooting.

### Directions

Answer the following questions.

1. Name the three categories of failures.

   a. ______________________

   b. ______________________

   c. ______________________

2. A customer comes into the shop with a constant failure. You have verified the failure. What is the next step?

   a. Repair the problem.

   b. Consult the service manual troubleshooting flowchart.

   c. Call the help line.

   d. Isolate the problem.

3. The service manager gives you a repair order that says a vehicle will not start. Where should troubleshooting the problem begin?
   a. Do a complete ignition system diagnosis.
   b. Remove and clean the carburetors.
   c. Check the engine stop switch and the fuel tank.
   d. Check the compression and leakdown.

4. A customer brings a 1987 Goldwing into a Honda service department with a dead battery. The technician has seen this problem before. A continuity test shows the stator is bad so he replaces the stator, charges the battery, flags the ticket, and calls the customer to pick it up. The customer calls the next day and says that the battery is dead again. What basic troubleshooting step did the technician neglect to perform?
   a. The battery was not replaced.
   b. A test ride was not made.
   c. The repair was not verified.
   d. The rectifier was not replaced.

5. A customer comes into the shop with a motorcycle that runs poorly at high speeds. She tells you that she has had this vehicle at three shops previously and they all cleaned and re-jetted the carburetors. Where do you start the troubleshooting procedure?
   a. Remove the carburetors to see if they really are clean inside.
   b. Remove the carburetors to check the jets.
   c. Call the other shops to see if they have any additional ideas.
   d. Assume the carburetors are clean the jetting is correct, and begin isolating the problem.

6. A technician is working on an ignition system with no spark. He replaces the plugs, battery, and the ignition coil but still no spark. He then pulls a new bike off the showroom floor, removes its ignition system, and puts it on the bike but to no avail. What mistake did the technician make in troubleshooting?
   a. He changed too many things at once.
   b. He assumed the ignition system was at fault without properly isolating the problem.
   c. He did not think out the possibilities before doing something drastic.
   d. All of the above.

7. The most difficult problems to troubleshoot are:
   a. Electrical
   b. Shifting
   c. Fuel related
   d. Handling related

8. Most electrical problems are related to the:
   a. Wiring connections
   b. Charging system
   c. Ignition system
   d. Fuel injection system

9. You suspect that a "no spark" condition is caused by the engine stop switch. How should you verify your suspicion?
   a. Replace it with a known good switch
   b. Order a new switch and replace it
   c. Check it with an ohm meter
   d. Cycle the switch off and on repeatedly

10. A technician replaces the rear tire on a 1999 Suzuki Katana. The customer returns the next day and complains that now the bike wobbles at speeds over 70 mph. This has never been a problem before. What is the logical cause of this problem?
    a. The steering head bearings are loose.
    b. The customer is trying to get the rear tire installed for free.
    c. The Katana has a design flaw.
    d. The rear tire is causing the problem.

11. The most difficult electrical problems to troubleshoot are ______________________________ problems.
    a. Ignition system
    b. Intermittent
    c. Fuel injection
    d. Constant

# Job Sheet 20-1

Name ______________________ Date __________ Instructor ______________

## Troubleshoot "No Start"

### Objective

After completing this job sheet in less than one hour, you will have demonstrated your mastery of the troubleshooting procedures necessary to diagnose the customer's contention.

### Directions

Answer the following questions.

### Tools and Equipment

Various hand tools

1. Customer's contention: Bike will not start.
2. Write down the troubleshooting steps used during your inspection of this problem.
   ____________________________________________
   ____________________________________________
   ____________________________________________
3. Explain how to correct the problem.
   ____________________________________________

INSTRUCTOR VERIFICATION:

# Job Sheet 20-2

Name ______________________________ Date __________ Instructor ______________

## Troubleshoot "Will Not Turn Over"

### Objective

After completing this job sheet in less than one hour, you will have demonstrated your mastery of the troubleshooting procedures necessary to diagnose the customer's contention.

### Directions

Answer the following questions.

### Tools and Equipment

Various hand tools, multimeter

1. Customer's contention: Bike will not turn over; battery just replaced.

2. Write down the troubleshooting steps used during your inspection of this problem.

   ______________________________________________________________

   ______________________________________________________________

   ______________________________________________________________

3. Explain how to correct the problem.

   ______________________________________________________________

INSTRUCTOR VERIFICATION:

# Job Sheet 20-3

Name ______________________ Date __________ Instructor ______________

## Troubleshoot "Bike Overheats"

### Objective

After completing this job sheet in less than one hour, you will have demonstrated your mastery of the troubleshooting procedures necessary to diagnose the customer's contention.

### Directions

Answer the following questions.

### Tools and Equipment

Various hand tools, multimeter

1. Customer's contention: Bike overheats in stop-and-go traffic.
2. Write down the troubleshooting steps used during your inspection of this problem.
   ________________________________________
   ________________________________________
   ________________________________________
3. Explain how to correct the problem.
   ________________________________________

INSTRUCTOR VERIFICATION:

# Job Sheet 20-4

Name ______________________ Date __________ Instructor ______________

## Troubleshoot "Bike Smokes Excessively"

### Objective

After completing this job sheet in less than one hour, you will have demonstrated your mastery of the troubleshooting procedures necessary to diagnose the customer's contention.

### Directions

Answer the following questions.

### Tools and Equipment

Various hand tools

1. Customer's contention: Bike smokes excessively.
2. Write down the troubleshooting steps used during your inspection of this problem.

   ______________________________________________

   ______________________________________________

   ______________________________________________

3. Explain how to correct the problem.

   ______________________________________________

INSTRUCTOR VERIFICATION:

# Job Sheet 20-5

Name ______________________________ Date __________ Instructor ______________

## Troubleshoot "Bike Will Not Idle"

### Objective

After completing this job sheet in less than one hour, you will have demonstrated your mastery of the troubleshooting procedures necessary to diagnose the customer's contention.

### Directions

Answer the following questions.

### Tools and Equipment

Various hand tools

1. Customer's contention: Bike will not idle after sitting all winter.
2. Write down the troubleshooting steps used during your inspection of this problem.
______________________________________________________________
______________________________________________________________
______________________________________________________________
3. Explain how to correct the problem.
______________________________________________________________

| INSTRUCTOR VERIFICATION: |
| --- |

# Job Sheet 20-6

Name ______________________________ Date __________ Instructor ______________

## Troubleshoot "Bike Jumps Out of Second Gear"

### Objective

After completing this job sheet in less than one hour, you will have demonstrated your mastery of the troubleshooting procedures necessary to diagnose the customer's contention.

### Directions

Answer the following questions.

### Tools and Equipment

Various hand tools

1. Customer's contention: Bike jumps out of second gear.

2. Write down the troubleshooting steps used during your inspection of this problem.
 ______________________________________________________________
 ______________________________________________________________
 ______________________________________________________________

3. Explain how to correct the problem.
 ______________________________________________________________

INSTRUCTOR VERIFICATION:

# Job Sheet 20-7

Name ______________________ Date __________ Instructor ____________

## Troubleshoot "FI Light On"

### Objective

After completing this job sheet in less than one hour, you will have demonstrated your mastery of the troubleshooting procedures necessary to diagnose the customer's contention.

### Directions

Answer the following questions.

### Tools and Equipment

Various hand tools

1. Customer's contention: FI light on.
2. Write down the troubleshooting steps used during your inspection of this problem.

   ______________________________________________

   ______________________________________________

   ______________________________________________

3. Explain how to correct the problem.

   ______________________________________________

INSTRUCTOR VERIFICATION:

# Job Sheet 20-7

Name ______________________ Date ______________ Instructor ______________________

## Troubleshoot "FI Light On"

### Objective

After completing this job sheet in less than one hour, you will have demonstrated your mastery of the troubleshooting procedures necessary to diagnose the customer's contention.

### Directions

Answer the following questions.

### Tools and Equipment

Various hand tools

1. Customer's contention: FI light on.

2. Write down the troubleshooting steps used during your inspection of this problem.

________________________________________________________________

________________________________________________________________

________________________________________________________________

3. Explain how to correct the problem.

________________________________________________________________

INSTRUCTOR VERIFICATION:

# Job Sheet 20-8

Name ______________________ Date __________ Instructor ______________

## Troubleshoot "FI Light Came On"

### Objective

After completing this job sheet in less than one hour, you will have demonstrated your mastery of the troubleshooting procedures necessary to diagnose the customer's contention.

### Directions

Answer the following questions.

### Tools and Equipment

Various hand tools

1. Customer's contention: FI light came on.
2. Write down the troubleshooting steps used during your inspection of this problem.
   ______________________________________________
   ______________________________________________
   ______________________________________________
3. Explain how to correct the problem.
   ______________________________________________

INSTRUCTOR VERIFICATION:

# Job Sheet 20-9

Name ______________________ Date __________ Instructor ______________

## Troubleshoot "Battery Goes Dead"

### Objective

After completing this job sheet in less than one hour, you will have demonstrated your mastery of the troubleshooting procedures necessary to diagnose the customer's contention.

### Directions

Answer the following questions.

### Tools and Equipment

Various hand tools, multimeter

1. Customer's contention: Battery goes dead.
2. Write down the troubleshooting steps used during your inspection of this problem.

   ______________________________________________

   ______________________________________________

   ______________________________________________

3. Explain how to correct the problem.

   ______________________________________________

INSTRUCTOR VERIFICATION:

# Job Sheet 20-10

Name ______________________ Date __________ Instructor ______________

## Troubleshoot "No Spark"

### Objective

After completing this job sheet in less than one hour, you will have demonstrated your mastery of the troubleshooting procedures necessary to diagnose the customer's contention.

### Directions

Answer the following questions.

### Tools and Equipment

Various hand tools, multimeter

1. Customer's contention: No spark.
2. Write down the troubleshooting steps used during your inspection of this problem.

   ______________________________________________

   ______________________________________________

   ______________________________________________

3. Explain how to correct the problem.

   ______________________________________________

INSTRUCTOR VERIFICATION:

# Job Sheet 20-11

Name ______________________________ Date __________ Instructor ______________

## Troubleshoot "Dim Headlight"

### Objective

After completing this job sheet in less than one hour, you will have demonstrated your mastery of the troubleshooting procedures necessary to diagnose the customer's contention.

### Directions

Answer the following questions.

### Tools and Equipment

Various hand tools, multimeter

1. Customer's contention: Dim headlight.
2. Write down the troubleshooting steps used during your inspection of this problem.
   ______________________________________________
   ______________________________________________
   ______________________________________________
3. Explain how to correct the problem.
   ______________________________________________

INSTRUCTOR VERIFICATION:

# Job Sheet 20-12

Name ______________________ Date __________ Instructor ______________

## Troubleshoot "Abnormal Noise"

### Objective
After completing this job sheet in less than one hour, you will have demonstrated your mastery of the troubleshooting procedures necessary to diagnose the customer's contention.

### Directions
Answer the following questions.

### Tools and Equipment
Various hand tools

1. Customer's contention: Abnormal noise.
2. Write down the troubleshooting steps used during your inspection of this problem.
   ______________________________________________
   ______________________________________________
   ______________________________________________
3. Explain how to correct the problem.
   ______________________________________________

INSTRUCTOR VERIFICATION:

# Job Sheet 20-13

Name ________________________ Date __________ Instructor ______________

## Troubleshoot "Will Not Take Throttle"

### Objective

After completing this job sheet in less than one hour, you will have demonstrated your mastery of the troubleshooting procedures necessary to diagnose the customer's contention.

### Directions

Answer the following questions.

### Tools and Equipment

Various hand tools

1. Customer's contention: Will not take throttle.
2. Write down the troubleshooting steps used during your inspection of this problem.
   ________________________________________
   ________________________________________
   ________________________________________
3. Explain how to correct the problem.
   ________________________________________

INSTRUCTOR VERIFICATION:

# Notes

# Notes

# Notes

# Notes